Preface

Welcome to the fascinating world of Exponential and Logarithmic Functions, a little book designed to guide high school and early college students. This book is not just a collection of formulas and theorems, but a journey into the heart of mathematical thinking, where we will explore the beauty, elegance, and practicality of exponential and logarithmic functions.

Exponential and logarithmic functions are fundamental to many areas of science, engineering, and economics. They are the backbone of concepts such as compound interest, population growth, radioactive decay, and the behavior of viruses, to name just a few. Understanding these functions is not only crucial for your mathematical education but also for comprehending the world around us.

This book is designed to make these complex concepts accessible and engaging. We present the material in a clear, concise manner, with a focus on understanding rather than memorizing. Each chapter begins with an introduction to the topic, followed by detailed explanations, and examples designed to reinforce your understanding.

We understand that every student has a unique learning style, and we have therefore included a variety of teaching tools, including diagrams, graphs, and step-by-step problem-solving strategies. We also encourage active learning, with practice problems that will challenge you to apply what you've learned in new and interesting ways.

Remember, mathematics is not a spectator sport. It requires active participation, curiosity, and a willingness to grapple with challenging problems. It's okay to struggle, to make mistakes, and to feel confused at times. These are all part of the learning process. What's important is to keep an open mind, to ask questions, and to never stop learning.

We hope that this book will not only help you succeed in your coursework but also inspire a lifelong appreciation for the beauty and power of mathematics. So, let's embark on this journey together, exploring the intriguing world of exponential and logarithmic functions.

Welcome to the adventure!

Table of Contents

1 What Is An Exponent?   3
   1.1 Definition 3
   1.2 How To Read a^n 5
   1.3 Properties of Exponents: 6
   1.4 Fractional Exponents 10

2 Exponential Functions   13
   2.1 Exponential Functions 13
     2.1.1 Definition 13
     2.1.2 Graphs of Exponential Functions 17
   2.2 Existence of Roots of Exponential Equations 20
   2.3 How to Solve an Exponential Equation 22
   2.4 Exponential Inequality 39

3 Logarithmic Functions   47
   3.1 Definition 47
   3.2 Logarithmic Functions 48
     3.2.1 Propertiies of Logarithm 49
     3.2.2 Graphs of Logarithmic Functions 59
   3.3 Logarithmic Equations and Inequalities: 62
     3.3.1 Logarithmic Equations 62
     3.3.2 Existence of Roots of Logarithmic Equations   62
     3.3.3 How To Solve It 63
   3.4 Logarithmic Inequalities 73

4 Problems   83

5 Solutions   93

Table of Contents

Chapter 1

What Is An Exponent?

Exponent plays an important role in Mathematics. It appears almost every context in Mathematics. We begin this chapter by introducing the definition of exponent.

1.1 Definition

> **Definition 1**
> An exponent refers to a number of times that a number is multiplied by itself.

> **Example 1**
> 1. The exponent of 2 in the product $2 \times 2 \times 2$ is 3.
> 2. The exponent of 3 in the product $3 \times 3 \times 3 \times 3 \times 3$ is 5.
> 3. The exponent of 5 in the product $5 \times 5 \times 5 \times 5$ is 4.

> **Definition 2**
> a to the nth power is the product of a by itself n times. To be convenient, we represent a to the nth power by
> $$a^n = \underbrace{a \times a \times ... \times a}_{n \text{ times of } a}.$$

Chapter 1. What Is An Exponent?

Below are elements of a power:

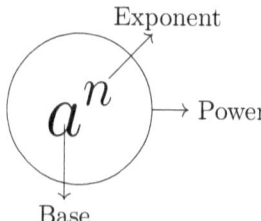

Example 2

Write the following products into exponential forms:

1. $2 \times 2 \times 2 = 2^3$
2. $3 \times 3 = 3^2$
3. $4 \times 4 \times 4 \times 4 = 4^4$
4. $(-1)(-1)(-1) = (-1)^3$
5. $\left(\dfrac{1}{3}\right)\left(\dfrac{1}{3}\right)\left(\dfrac{1}{3}\right)\left(\dfrac{1}{3}\right) = \left(\dfrac{1}{3}\right)^4$
6. $\left(\dfrac{2}{5}\right)\left(\dfrac{2}{5}\right)\left(\dfrac{2}{5}\right) = \left(\dfrac{2}{5}\right)^3$

Example 3

Compute the following expressions:

1. 3^2
2. 5^3
3. 6^4
4. 1^{10}
5. 10^3
6. $\left(\dfrac{1}{2}\right)^5$
7. $\left(-\dfrac{1}{3}\right)^3$
8. $\left(\dfrac{3}{2}\right)^2$
9. $\left(-\dfrac{5}{3}\right)^3$
10. $\left(-\dfrac{1}{4}\right)^3$

Solution.

1. $3^2 = 3 \times 3 = 9$

1.2. How To Read a^n

2. $5^3 = 5 \times 5 \times 5 = 125$

3. $6^4 = 6 \times 6 \times 6 \times 6 = 1296$

4. $1^{10} = 1 \times 1 \times 1 \times 1 \times 1 \times 1 \times 1 \times 1 \times 1 \times 1 = 1$

5. $10^3 = 10 \times 10 \times 10 = 1000$

6. $\left(\dfrac{1}{2}\right)^5 = \dfrac{1}{2} \times \dfrac{1}{2} \times \dfrac{1}{2} \times \dfrac{1}{2} \times \dfrac{1}{2} = \dfrac{1}{32}$

7. $\left(-\dfrac{1}{3}\right)^3 = \left(-\dfrac{1}{3}\right)\left(-\dfrac{1}{3}\right)\left(-\dfrac{1}{3}\right) = -\dfrac{1}{27}$

8. $\left(\dfrac{3}{2}\right)^2 = \left(\dfrac{3}{2}\right)\left(\dfrac{3}{2}\right) = \dfrac{9}{4}$

9. $\left(-\dfrac{5}{3}\right)^3 = \left(-\dfrac{5}{3}\right)\left(-\dfrac{5}{3}\right)\left(-\dfrac{5}{3}\right) = -\dfrac{125}{27}$

10. $\left(-\dfrac{1}{4}\right)^3 = \left(-\dfrac{1}{4}\right)\left(-\dfrac{1}{4}\right)\left(-\dfrac{1}{4}\right) = -\dfrac{1}{64}$

1.2 How To Read a^n

We already introduced readers the definition of the nth power of a real number. In this section, readers will learn how to read a^n.

- 2^1 read as 2 to the first power;
- 2^2 read as 2 to the second power 2 or 2 squared;
- 2^3 read as 2 to the third power 3 or 2 cubed;
- 2^4 read as 2 to the fourth power 4;
- 2^n read as 2 to the nth power.

> **Example 4**
>
> Write the following powers in words:
>
> 1. 2^5 3. 4^3 5. 9^5
>
> 2. 3^7 4. 5^2 6. 7^{10}

Solution.

1. 2^5: 2 to the fifth power;
2. 3^7: 3 to the seventh power;
3. 4^3: 4 to the third power or 4 cubed;
4. 5^2: 5 to the second power or 5 squared;
5. 9^5: 9 to the fifth power;
6. 7^{10}: 7 to the tenth power.

> **Note 1**
> We prefer reading a^2 as a squared rather than a to the second power. We prefer reading a^3 as a cubed rather than a to the third power.

1.3 Properties of Exponents:

> **General 1**
> 1. $x^m \times x^n = x^{m+n}$;
> 2. $\dfrac{x^m}{x^n} = x^{m-n}$ for all $x \neq 0$;
> 3. $x^{-n} = \dfrac{1}{x^n}$ for all $x \neq 0$;
> 4. $x^0 = 1$ for all $x \neq 0$;
> 5. $(xy)^n = x^n y^n$;
> 6. $\left(\dfrac{x}{y}\right)^n = \dfrac{x^n}{y^n}$ for all $y \neq 0$.

Proof. 1. $x^m \times x^n = x^{m+n}$

By the definition of exponent, we have

$$x^m \times x^n = \underbrace{x \times x \times \ldots \times x}_{n \text{ times of } x} \times \underbrace{x \times x \times \ldots \times x}_{m \text{ times of } x}$$

1.3. Properties of Exponents:

$$= \underbrace{x \times x \times \ldots \times x}_{m+n \text{ times of } x}$$
$$= x^{m+n}.$$

Below are some examples of the property: $x^m \times x^n = x^{m+n}$.

Example 5

(a) $3^4 \times 3^5 = 3^{4+5} = 3^9$

(b) $4^2 \times 4^7 = 4^{2+7} = 4^9$

(c) $5^4 \times 5^6 = 5^{4+6} = 5^{10}$

2. $\dfrac{x^m}{x^n} = x^{m-n}$ for all $x \neq 0$

For $m > n$, we obtain

$$\dfrac{x^m}{x^n} = \dfrac{\underbrace{x \times x \times \ldots \times x}_{m \text{ times of } x}}{\underbrace{x \times x \times \ldots \times x}_{n \text{ times of } x}}$$

$$= \dfrac{\underbrace{x \times x \times \ldots \times x}_{m-n \text{ times of } x} \times \underbrace{x \times x \times \ldots \times x}_{n \text{ times of } x}}{\underbrace{x \times x \times \ldots \times x}_{n \text{ times of } x}}$$

$$= \underbrace{x \times x \times \ldots \times x}_{m-n \text{ times of } x}.$$

In general, we define $\dfrac{x^m}{x^n} = x^{m-n}$ for the case $m = n$ and $m < n$. We illustrate the above property by the following examples:

Example 6

(a) $\dfrac{3^4}{3} = 3^{4-1} = 3^3$

(b) $\dfrac{4^5}{4^2} = 4^{5-2} = 4^3$

(c) $\dfrac{7^3}{7^2} = 7^{3-2} = 7$

(d) $\dfrac{9^{10}}{9^4} = 9^{10-4} = 9^6$

3. $x^0 = 1$ for all $x \neq 0$

For all $x \neq 0$, we have $\dfrac{x^m}{x^n} = x^{m-n}$.

For $m = n$, we obtain $\dfrac{x^n}{x^n} = x^{n-n}$. Then $1 = x^0$.

Therefore, $x^0 = 1$ for all $x \neq 0$.

Example 7

(a) $1^0 = 1$

(b) $2^0 = 1$

(c) $2000^0 = 1$

(d) $10000000^0 = 1$

4. $x^{-n} = \dfrac{1}{x^n}$ for all $x \neq 0$

Using property 2, we have $\dfrac{x^m}{x^n} = x^{m-n}$ for all $x \neq 0$.

For $m = 0$, it follows that $\dfrac{x^0}{x^n} = x^{0-n}$. Then $\dfrac{1}{x^n} = x^{-n}$.

Therefore, $x^{-n} = \dfrac{1}{x^n}$ for all $x \neq 0$.

Example 8

(a) $2^{-2} = \dfrac{1}{2^2} = \dfrac{1}{4}$

(b) $3^{-2} = \dfrac{1}{3^2} = \dfrac{1}{9}$

(c) $4^{-2} = \dfrac{1}{4^2} = \dfrac{1}{16}$

1.3. Properties of Exponents:

> (d) $6^{-2} = \dfrac{1}{6^2} = \dfrac{1}{36}$

5. $(xy)^n = x^n y^n$
By the definition, we have
$$(xy)^n = \underbrace{(xy)(xy)(xy)\ldots(xy)}_{n \text{ times of } xy}$$
$$= \underbrace{x \times x \times \ldots \times x}_{n \text{ times of } x} \times \underbrace{y \times y \times \ldots \times y}_{n \text{ times of } y}$$
$$= x^n y^n.$$

Therefore, $(xy)^n = x^n y^n.$

Example 9

(a) $2^2 \times 3^2 = (2 \times 3)^2 = 6^2 = 36$

(b) $3^2 \times 4^2 = (3 \times 4)^2 = 12^2 = 144$

(c) $5^2 \times 2^2 = (5 \times 2)^2 = 10^2 = 100$

(d) $\left(\dfrac{1}{2}\right)^2 \times \left(\dfrac{1}{4}\right)^2 = \left(\dfrac{1}{2} \times \dfrac{1}{4}\right)^2 = \left(\dfrac{1}{8}\right)^2 = \dfrac{1}{64}$

6. $\left(\dfrac{x}{y}\right)^n = \dfrac{x^n}{y^n}$ for all $y \neq 0$
By the definition of exponent, we have
$$\left(\dfrac{x}{y}\right)^n = \underbrace{\dfrac{x}{y} \times \dfrac{x}{y} \times \ldots \times \dfrac{x}{y}}_{n \text{ times of } \frac{x}{y}}$$
$$= \dfrac{\underbrace{x \times x \times \ldots \times x}_{n \text{ times of } x}}{\underbrace{y \times y \times \ldots \times y}_{n \text{ times of } y}}$$
$$= \dfrac{x^n}{y^n}.$$

Therefore, $\left(\dfrac{x}{y}\right)^n = \dfrac{x^n}{y^n}$ for all $y \neq 0$.

Example 10

(a) $\dfrac{2^5}{4^5} = \left(\dfrac{2}{4}\right)^5 = \left(\dfrac{1}{2}\right)^5 = \dfrac{1}{32}$

(b) $\dfrac{3^2}{27^2} = \left(\dfrac{3}{27}\right)^2 = \left(\dfrac{1}{9}\right)^2 = \dfrac{1}{81}$

(c) $\dfrac{5^3}{10^3} = \left(\dfrac{5}{10}\right)^3 = \left(\dfrac{1}{2}\right)^3 = \dfrac{1}{8}$

(d) $\dfrac{2^4}{16^2} = \dfrac{2^4}{(4^2)^2} = \dfrac{2^4}{4^4} = \left(\dfrac{2}{4}\right)^4 = \left(\dfrac{1}{2}\right)^4 = \dfrac{1}{16}$

□

Let us now observe the values of the nth power of -1. We have $(-1)^1 = -1, (-1)^2 = 1, (-1)^3 = -1, \ldots$ From the above computation, we obtain the following results:

General 2

- If n is an even number, then $(-1)^n = 1$.
- If n is an odd number, then $(-1)^n = -1$.

1.4 Fractional Exponents

For all $a > 0$, we obtain $a^{\frac{m}{n}} = \sqrt[n]{\left(a^{\frac{m}{n}}\right)^n} = \sqrt[n]{a^{\frac{m}{n} \times n}} = \sqrt[n]{a^m}$.

Theorem 1

For all $a > 0$, we obtain $\sqrt[n]{a^m} = a^{\frac{m}{n}}$.

Example 11

Write the following numbers as fractional exponents:

1.4. Fractional Exponents

1. $\sqrt{2}$
2. $\sqrt{3^7}$
3. $\sqrt[3]{7^2}$
4. $\sqrt[3]{5^2}$
5. $\sqrt[4]{31}$
6. $\dfrac{1}{\sqrt{5}}$
7. $\dfrac{3}{\sqrt[3]{3}}$
8. $\dfrac{7}{\sqrt[5]{7^2}}$
9. $\dfrac{4}{\sqrt[4]{2}}$
10. $\dfrac{\sqrt{2}}{\sqrt[3]{2}}$

Solution.

1. $\sqrt{2} = 2^{\frac{1}{2}}$
2. $\sqrt{3^7} = 3^{\frac{7}{2}}$
3. $\sqrt[3]{7^2} = 7^{\frac{2}{3}}$
4. $\sqrt[3]{5^2} = 5^{\frac{2}{3}}$
5. $\sqrt[4]{31} = 31^{\frac{1}{4}}$
6. $\dfrac{1}{\sqrt{5}} = \dfrac{1}{5^{\frac{1}{2}}} = 5^{-\frac{1}{2}}$
7. $\dfrac{3}{\sqrt[3]{3}} = \dfrac{\sqrt[3]{3^3}}{\sqrt[3]{3}} = \sqrt[3]{3^2} = 3^{\frac{2}{3}}$
8. $\dfrac{7}{\sqrt[5]{7^2}} = \dfrac{\sqrt[5]{7^5}}{\sqrt[5]{7^2}} = \sqrt[5]{7^3} = 7^{\frac{3}{5}}$
9. $\dfrac{4}{\sqrt[4]{2}} = \dfrac{2^2}{\sqrt[4]{2}} = \dfrac{2^2}{2^{\frac{1}{4}}} = 2^{2-\frac{1}{4}} = 2^{\frac{8-1}{4}} = 2^{\frac{7}{4}}$
10. $\dfrac{\sqrt{2}}{\sqrt[3]{2}} = \dfrac{2^{\frac{1}{2}}}{2^{\frac{1}{3}}} = 2^{\frac{1}{2}-\frac{1}{3}} = 2^{\frac{3-2}{6}} = 2^{\frac{1}{6}}$

Chapter 1. What Is An Exponent?

Chapter 2

Exponential Functions

Understanding exponential functions is key to exploring more complex mathematical concepts and applying them to solve practical problems. They are a fundamental tool in the mathematical toolkit, providing a powerful way to represent growth and decay processes that occur in the natural and social sciences. We start this chapter by discussing the definition of exponential functions.

2.1 Exponential Functions

2.1.1 Definition

> **Definition 3**
> An exponential function f is a function from $\mathbb{R}$ to $\mathbb{R}^+$ which is defined by $f(x) = a^x$, where $a > 0$ and $a \neq 1$.

> **Example 12**
> Below are some examples of exponential functions:
>
> - $f(x) = 2^x$
> - $g(x) = \left(\dfrac{1}{2}\right)^x$
> - $h(x) = 0.6^x$

Chapter 2. Exponential Functions

> **Example 13**
>
> Suppose that f is an exponential function. Prove that
>
> 1. $f(0) = 1$;
> 2. $f(x+y) = f(x)f(y)$;
> 3. $f(x-y) = \dfrac{f(x)}{f(y)}$;
> 4. $f(x) + f(-x) \geq 2$.

Solution. Prove that

1. $f(0) = 1$
   Since f is an exponential function, then $f(x) = a^x$, where $a > 0$ and $a \neq 1$. It follows that $f(0) = a^0 = 1$.
   $\boxed{\text{Therefore, } f(0) = 1.}$

2. $f(x+y) = f(x)f(y)$
   We have $f(x+y) = a^{x+y} = a^x \times a^y = f(x) \times f(y)$.
   $\boxed{\text{Therefore, } f(x+y) = f(x)f(y).}$

3. $f(x-y) = \dfrac{f(x)}{f(y)}$
   We have $f(x-y) = a^{x-y} = \dfrac{a^x}{a^y} = \dfrac{f(x)}{f(y)}$.
   $\boxed{\text{Therefore, } f(x-y) = \dfrac{f(x)}{f(y)}.}$

4. $f(x) + f(-x) \geq 2$
   We have
   $$f(x) + f(-x)$$
   $$= a^x + a^{-x}$$
   $$= a^x + \dfrac{1}{a^x}$$
   $$= \left(\sqrt{a^x}\right)^2 - 2\left(\sqrt{a^x}\right)\left(\sqrt{\dfrac{1}{a^x}}\right) + \left(\sqrt{\dfrac{1}{a^x}}\right)^2 + 2\left(\sqrt{a^x}\right)\left(\sqrt{\dfrac{1}{a^x}}\right)$$
   $$= \left(\sqrt{a^x} - \sqrt{\dfrac{1}{a^x}}\right)^2 + 2 \geq 2.$$

   $\boxed{\text{Consequently, } f(x) + f(-x) \geq 2.}$

2.1. Exponential Functions

Remark 1. A square is always nonnegative. That is, for all real numbers x, we obtain $x^2 \geq 0$.

> **Example 14**
>
> Given a function f which is defined by $f(x) = \dfrac{1}{1+a^{2x}}$, $a > 0$. Prove that $f(x) + f(-x) = 1$.

Solution. We have $f(x) = \dfrac{1}{1+a^{2x}}$, where $a > 0$.

Substitute x by $-x$, we obtain $f(-x) = \dfrac{1}{1+a^{-2x}}$.

It follows that

$$\begin{aligned} f(x) + f(-x) &= \frac{1}{1+a^{2x}} + \frac{1}{1+a^{-2x}} \\ &= \frac{1}{1+a^{2x}} + \frac{1}{1+\frac{1}{a^{2x}}} \\ &= \frac{1}{1+a^{2x}} + \frac{a^{2x}}{a^{2x}+1} \\ &= \frac{1}{1+a^{2x}} + \frac{a^{2x}}{1+a^{2x}} \\ &= \frac{1+a^{2x}}{1+a^{2x}} = 1. \end{aligned}$$

Therefore, $f(x) + f(-x) = 1.$

> **Example 15**
>
> Given a function f which is defined by $f(x) = \dfrac{2}{4^x + 2}$.
> Evaluate $f\left(\dfrac{1}{n}\right) + f\left(\dfrac{2}{n}\right) + \ldots + f\left(\dfrac{n-1}{n}\right)$.

Solution. We have $f(x) = \dfrac{2}{4^x + 2}$. Then

$$f(1-x) = \frac{2}{4^{1-x} + 2}$$

$$= \frac{2}{\frac{4}{4^x}+2}$$

$$= \frac{2\times 4^x}{4+2\times 4^x}$$

$$= \frac{2\times 4^x}{2(2+4^x)}$$

$$= \frac{4^x}{4^x+2}.$$

It follows that

$$f(x)+f(1-x) = \frac{2}{4^x+2} + \frac{4^x}{4^x+2} = \frac{4^x+2}{4^x+2} = 1.$$

Let

$$S = f\left(\frac{1}{n}\right) + f\left(\frac{2}{n}\right) + \ldots + f\left(\frac{n-1}{n}\right). \qquad (1)$$

It implies that

$$S = f\left(\frac{n-1}{n}\right) + f\left(\frac{n-2}{n}\right) + \ldots + f\left(\frac{2}{n}\right) + f\left(\frac{1}{n}\right). \qquad (2)$$

Adding (1) and (2), we obtain

$$2S = \left[f\left(\frac{1}{n}\right) + f\left(\frac{n-1}{n}\right)\right] + \left[f\left(\frac{2}{n}\right) + f\left(\frac{n-2}{n}\right)\right] + \ldots$$
$$+ \left[f\left(\frac{n-1}{n}\right) + f\left(\frac{1}{n}\right)\right]$$
$$= \left[f\left(\frac{1}{n}\right) + f\left(1-\frac{1}{n}\right)\right] + \left[f\left(\frac{2}{n}\right) + f\left(1-\frac{2}{n}\right)\right] + \ldots$$
$$+ \left[f\left(\frac{n-1}{n}\right) + f\left(1-\frac{n-1}{n}\right)\right]$$
$$= \underbrace{1+1+\ldots+1}_{n \text{ terms}}$$
$$= n.$$

Therefore, $S = \dfrac{n}{2}.$

2.1. Exponential Functions

2.1.2 Graphs of Exponential Functions

In this part, we will introduce readers graphs of exponential functions and their properties. Below are examples of how to draw graphs of exponential functions.

> **Example 16**
>
> Draw the graphs of the following exponential functions:
>
> 1. $f(x) = 2^x$
>   To draw the graph of f, we first construct the following table of values for it:
>
> | x | -1 | 0 | 1 |
> |---|---|---|---|
> | y | $\dfrac{1}{2}$ | 1 | 2 |
>
>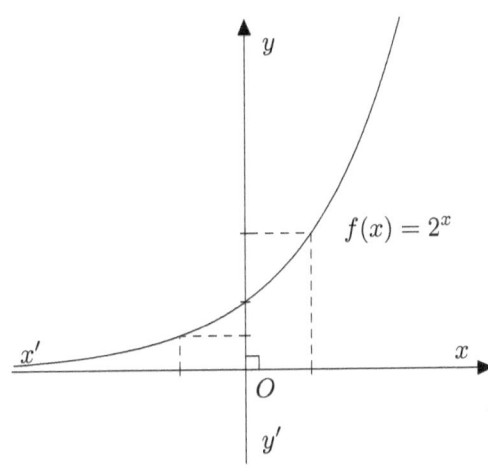
>
> 2. $f(x) = 3^x$
>   To draw the graph of f, we first construct the following table of values for it:
>
> | x | -1 | 0 | 1 |
> |---|---|---|---|
> | y | $\dfrac{1}{3}$ | 1 | 3 |

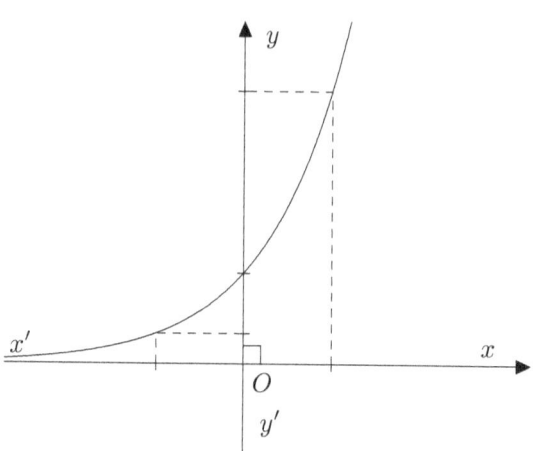

3. $f(x) = \left(\dfrac{1}{2}\right)^x$

To draw the graph of f, we first construct the following table of values for it:

x	-1	0	1
y	2	1	$\dfrac{1}{2}$

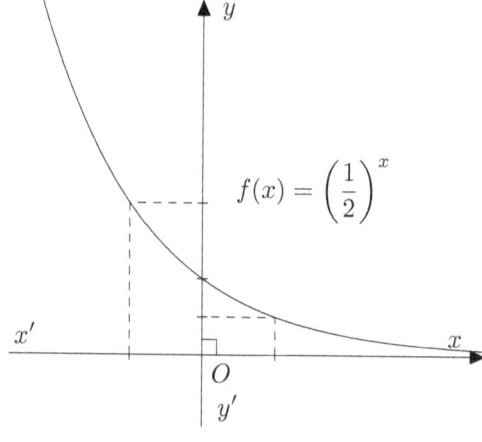

2.1. Exponential Functions

4. $f(x) = \left(\dfrac{1}{3}\right)^x$

To draw the graph of f, we first construct the following table of values for it:

x	-1	0	1
y	3	1	$\dfrac{1}{3}$

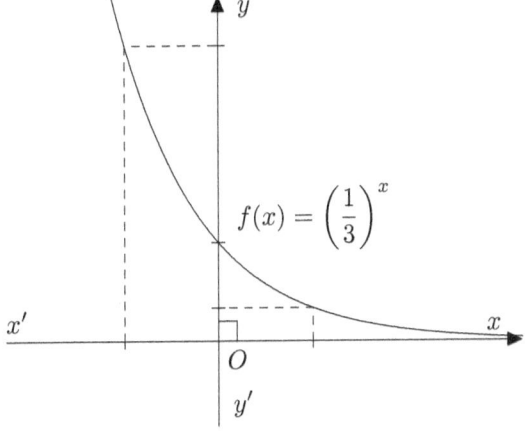

From the above examples, we see that the graph C of a function f which is defined by $f(x) = a^x$, where $a > 1$ has the following properties:

1. C passes through $(0, 1)$.

2. f has outputs for all $x \in \mathbb{R}$. That is, the function f is well-defined for all $x \in \mathbb{R}$.

3. The outputs of f are always positive. Namely, $f(x) > 0$ for all $x \in \mathbb{R}$.

4. The graph C is increasing from left to right.

5. C approaches x-axis as x tends to $-\infty$.

6. C is infinitely increasing as x tends to $+\infty$.

Conversely, we see that the graph C of a function f which is defined by $f(x) = a^x$, where $0 < a < 1$ has the following properties:

1. C passes through $(0, 1)$.

2. f has outputs for all $x \in \mathbb{R}$. That is, the function f is well-defined for all $x \in \mathbb{R}$.

3. The outputs of f are always positive. Namely, $f(x) > 0$ for all $x \in \mathbb{R}$.

4. The graph C is decreasing from left to right.

5. C approaches x-axis as x tends to $+\infty$.

6. C is infinitely increasing as x tends to $-\infty$.

> **General 3**
>
> Given an exponential function f which is defined by
> $$f : \mathbb{R} \to \mathbb{R}^+$$
> $$x \mapsto f(x) = a^x, \quad \text{where } a > 0 \text{ and } a \neq 1.$$
>
> - If $a > 1$, f is an increasing function.
> - If $0 < a < 1$, f is a decreasing function.

2.2 Existence of Roots of Exponential Equations

An exponential equation is an equation that can be written in the form of $a^x = b$, where $a > 0, b > 0$ and $a \neq 1$. The number of roots of the exponential equation $a^x = b$ is the number of points of intersections of the graph of $y = a^x$ and the line $y = b$. There are two cases to consider.

1. If $a > 1$, then $f(x) = a^x$ is an increasing function. In this case, we obtain the following graph:

2.2. Existence of Roots of Exponential Equations

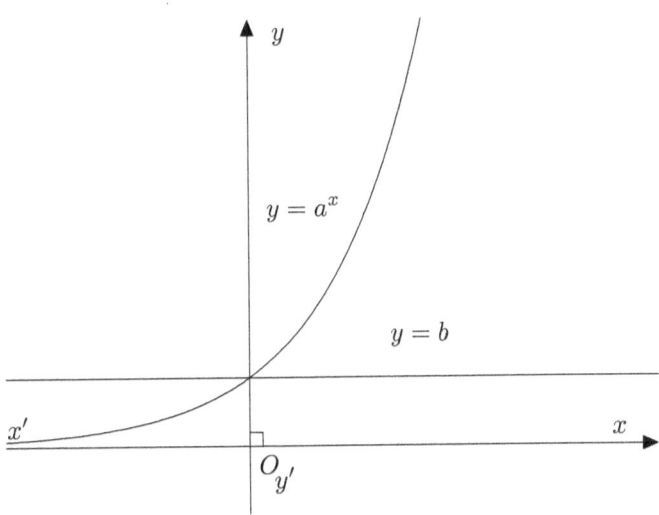

2. If $0 < a < 1$, then $f(x) = a^x$ is a decreasing function. In this case, we obtain the following graph:

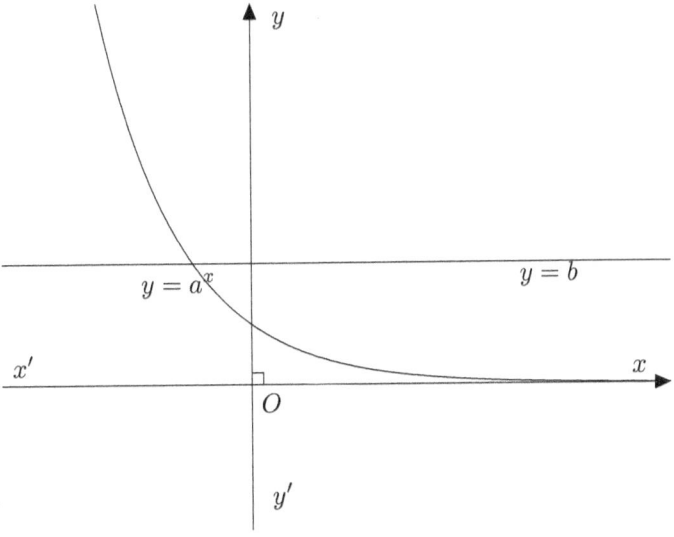

From the above graphs, we see that the curve $y = a^x$ and the line $y = b$ always intersect at a point. Therefore, the exponential equation $a^x = b$ has a unique solution for all $a, b > 0$.

2.3 How to Solve an Exponential Equation

Theorem 2
Suppose that $a > 0$ and $a \neq 1$. Then $a^x = a^y$ if and only if $x = y$.

To understand clearly, see the following example.

Example 17
Solve the following exponential equations:

1. $3^x = 3$
2. $5^x = 25$
3. $6^x = 36$
4. $6^{x+1} = 216$
5. $2 \times 2^{x+1} + 2^x = 20$

Solution. Solve the following exponential equations:

1. $3^x = 3$
   We have $3^x = 3$. Then $3^x = 3^1$. It follows that $x = 1$.
   Therefore, $x = 1$ is the solution of the equation.

2. $5^x = 25$
   We first write 25 as a power of 5. We have $5^x = 5^2$.
   Then $x = 2$.
   Therefore, $x = 2$ is the solution of the equation.

3. $6^x = 36$
   We first write 36 as a power of 6. We have $36 = 6^2$.
   Then $6^x = 6^2$.
   It follows that $x = 2$.
   Therefore, $x = 2$ is the solution of the equation.

4. $6^{x+1} = 216$
   We have $6^{x+1} = 216$. Then $6^{x+1} = 6^3$.
   It follows that $x + 1 = 3$. Then $x = 3 - 1 = 2$.
   Therefore, $x = 2$ is the solution of the equation.

2.3. How to Solve an Exponential Equation

5. $2 \times 2^{x+1} + 2^x = 20$
   We have $2 \times 2^{x+1} + 2^x = 20$. Then
   $$2 \times 2 \times 2^x + 2^x = 20$$
   or
   $$4 \times 2^x + 2^x = 20.$$
   We obtain $5 \times 2^x = 20$. Then $2^x = 4 = 2^2$.
   Therefore, $x = 2$ is the solution of the equation.

> **General 4**
>
> To solve an exponential equation $a^x = b$, we follow the following steps:
>
> 1. Write b as a power of a. Suppose that $b = a^y$.
> 2. From $a^x = a^y$, we obtain $x = y$.

> **Example 18**
>
> Solve the following exponential equations:
>
> 1. $2^x = 4$
> 2. $3^x = 81$
> 3. $7^x = 49$
> 4. $5^x = \dfrac{1}{25}$
> 5. $8^x = 2$
> 6. $9^x = \dfrac{1}{27}$

Solution. Solve the following exponential equations:

1. $2^x = 4$
   We have
   $$2^x = 4$$
   $$2^x = 2^2$$
   Therefore, $x = 2$ is the solution of the equation.

2. $3^x = 81$
   We have
   $$3^x = 81$$

$$3^x = 3^4$$

Therefore, $x = 4$ is the solution of the equation.

3. $7^x = 49$
   We have
   $$7^x = 49$$
   $$7^x = 7^2$$

Therefore, $x = 2$ is the solution of the equation.

4. $5^x = \dfrac{1}{25}$
   We have
   $$5^x = \frac{1}{25}$$
   $$5^x = \frac{1}{5^2}$$
   $$5^x = 5^{-2}$$

Therefore, $x = -2$ is the solution of the equation.

5. $8^x = 2$
   We have
   $$8^x = 2$$
   $$\left(2^3\right)^x = 2$$
   $$2^{3x} = 2$$

It follows that $3x = 1$.

Therefore, $x = \dfrac{1}{3}$ is the solution of the equation.

6. $9^x = \dfrac{1}{27}$
   We have
   $$9^x = \frac{1}{27}$$
   $$\left(3^2\right)^x = \frac{1}{3^3}$$

2.3. How to Solve an Exponential Equation

$$3^{2x} = 3^{-3}$$

It follows that $2x = -3$.

Therefore, $x = -\dfrac{3}{2}$ is the solution of the equation.

Example 19

Solve the following exponential equations:

1. $3^{x^2-3x+2} = 1$
2. $4^{x^2-x} = \dfrac{1}{16}$
3. $5^{x^3-2x} = \dfrac{1}{5}$
4. $2^{x^2+2x} = 3^{-x^2-2x}$

Solution. Solve the following exponential equations:

1. $3^{x^2-3x+2} = 1$
   We have
   $$3^{x^2-3x+2} = 1$$
   $$3^{x^2-3x+2} = 3^0$$

   Then $x^2 - 3x + 2 = 0$. It is trivial to see that the sum of all the coefficients of the quadratic equation is zero.
   Hence, $x_1 = 1$ and $x_2 = \dfrac{c}{a} = \dfrac{2}{1} = 2$.

   Therefore, $x \in \{1, 2\}$.

2. $4^{x^2-x} = \dfrac{1}{16}$
   We have
   $$4^{x^2-x} = \dfrac{1}{16}$$
   $$4^{x^2-x} = \dfrac{1}{4^2}$$
   $$4^{x^2-x} = 4^{-2}$$

   Then $x^2 - x = -2$. It follows that $x^2 - x + 2 = 0$.
   The discriminant of the quadratic equation is
   $$\Delta = b^2 - 4ac$$

$$= (-1)^2 - 4(1)(2)$$
$$= 1 - 8$$
$$= -7 < 0.$$

The quadratic equation has no roots.
Consequently, the equation has no solutions.

3. $5^{x^3-2x} = \dfrac{1}{5}$
   We have
   $$5^{x^2-2x} = \dfrac{1}{5}$$
   $$5^{x^2-2x} = 5^{-1}$$

   It follows that $x^2 - 2x = -1$. Then $x^2 - 2x + 1 = 0$.
   We obtain $(x-1)^2 = 0$. Consequently, $x - 1 = 0$.
   Therefore, $x = 1$.

4. $2^{x^2+2x} = 3^{-x^2-2x}$
   We have
   $$2^{x^2+2x} = 3^{-x^2-2x}$$
   $$2^{x^2+2x} = 3^{-(x^2+2x)}$$
   $$2^{x^2+2x} \times 3^{x^2+2x} = 1$$
   $$6^{x^2+2x} = 6^0$$

   It follows that $x^2 + 2x = 0$. Then $x(x+2) = 0$.
   Consequently, $\begin{bmatrix} x = 0 \\ x + 2 = 0 \end{bmatrix}$. Then $\begin{bmatrix} x = 0 \\ x = -2 \end{bmatrix}$.
   Therefore, $x \in \{-2, 0\}$.

Example 20

Solve the following exponential equations:

1. $3^x = \dfrac{1}{3}$

2. $3^{-x^2-2x} = \dfrac{1}{27}$

2.3. How to Solve an Exponential Equation

3. $2^x = \dfrac{1}{32}$

4. $\left(\dfrac{1}{2}\right)^x = 16$

5. $5 \times 2^{x+1} + 6 \times 2^{x+2} + 7 \times 2^{x+3} = 90$

Solution. Solve the following exponential equations:

1. $3^x = \dfrac{1}{3}$

   We have $3^x = \dfrac{1}{3}$. Then $3^x = 3^{-1}$.
   It implies that $x = -1$.

   Therefore, $x = -1$.

2. $3^{-x^2-2x} = \dfrac{1}{27}$

   We have $3^{-x^2-2x} = \dfrac{1}{27}$.
   Then
   $$3^{-x^2-2x} = \dfrac{1}{3^3}$$
   or
   $$3^{-x^2-2x} = 3^{-3}.$$
   It implies that $-x^2 - 2x = -3$. Then $-x^2 - 2x + 3 = 0$.
   Since the sum of all coefficients of the quadratic equation is zero, we obtain $x_1 = 1$ and $x_2 = \dfrac{c}{a} = \dfrac{3}{-1} = -3$.

   Therefore, $x \in \{-3, 1\}$.

3. $2^x = \dfrac{1}{32}$

   We have $2^x = \dfrac{1}{32}$. Then $2^x = \dfrac{1}{2^5}$. It follows that $2^x = 2^{-5}$.

   Therefore, $x = -5$.

4. $\left(\dfrac{1}{2}\right)^x = 16$

   We have $\left(\dfrac{1}{2}\right)^x = 16$. Then $\left(\dfrac{1}{2}\right)^x = 2^4$.

It follows that $\left(\dfrac{1}{2}\right)^x = \left(\dfrac{1}{2}\right)^{-4}$.

Therefore, $x = -4$.

5. $5 \times 2^{x+1} + 6 \times 2^{x+2} + 7 \times 2^{x+3} = 90$
We have $5 \times 2^{x+1} + 6 \times 2^{x+2} + 7 \times 2^{x+3} = 90$.
Then $5 \times 2 \times 2^x + 6 \times 2^2 \times 2^x + 7 \times 2^3 \times 2^x = 90$.
It follows that $10 \times 2^x + 24 \times 2^x + 56 \times 2^x = 90$.
Then $90 \times 2^x = 90$. We obtain $2^x = 1 = 2^0$.

Therefore, $x = 0$.

Example 21

Solve the following exponential equations:

1. $2^{2x} + 2^x - 2 = 0$

2. $5 \times 3^{2x} - 3^x - 4 = 0$

3. $2^x + 2^{-x} - \dfrac{5}{2} = 0$

4. $7^x + \dfrac{2}{7^x} = 3$

Solution. Solve the following exponential equations:

1. $2^{2x} + 2^x - 2 = 0$
Let $t = 2^x$. Then $t > 0$ for all $x \in \mathbb{R}$. The given equation can be written as
$$t^2 + t - 2 = 0.$$
Then $(t-1)(t+2) = 0$.
Since $t + 2 > 0$ for all $x \in \mathbb{R}$, it turns out that $t - 1 = 0$.
Then $t = 1$.
It follows that $2^x = 1$. Hence, $2^x = 2^0$.

Therefore, $x = 0$.

2. $5 \times 3^{2x} - 3^x - 4 = 0$
Let $t = 3^x$. We obtain $t > 0$ for all $x \in \mathbb{R}$. The given equation can be written as $5t^2 - t - 4 = 0$.
Then $(t-1)(5t+4) = 0$.
Since $5t + 4 > 0$ for all $x \in \mathbb{R}$, we obtain $t - 1 = 0$.
Then $t = 1$. It follows that $3^x = 1 = 3^0$.

Therefore, $x = 0$.

2.3. How to Solve an Exponential Equation

3. $2^x + 2^{-x} - \dfrac{5}{2} = 0$

    We have $2^x + 2^{-x} - \dfrac{5}{2} = 0$. Then $2^x + \dfrac{1}{2^x} - \dfrac{5}{2} = 0$.
    Let $t = 2^x$. Then $t > 0$ for all $x \in \mathbb{R}$.
    The given equation can be written as
    $$t + \dfrac{1}{t} - \dfrac{5}{2} = 0$$
    or
    $$2t^2 - 5t + 2 = 0.$$
    It follows that $(2t-1)(t-2) = 0$.
    Hence, $\left[\begin{array}{l} t - 2 = 0 \\ 2t - 1 = 0 \end{array}\right.$ or $\left[\begin{array}{l} t = 2 \\ t = \dfrac{1}{2} \end{array}\right.$.

    - If $t = 2$, we obtain $2^x = 2$. Then $x = 1$.
    - If $t = \dfrac{1}{2}$, we obtain $2^x = \dfrac{1}{2} = 2^{-1}$.

    Thus, $x = -1$.
    $\boxed{\text{Therefore, } x \in \{-1, 1\}.}$

4. $7^x + \dfrac{2}{7^x} = 3$

    Let $t = 7^x$. Then $t > 0$ for all $x \in \mathbb{R}$. The given equation can be written as $t + \dfrac{2}{t} = 3$. Then $t^2 - 3t + 2 = 0$. It implies that $(t-1)(t-2) = 0$.
    Hence, $\left[\begin{array}{l} t - 1 = 0 \\ t - 2 = 0 \end{array}\right.$ or $\left[\begin{array}{l} t = 1 \\ t = 2 \end{array}\right.$.

    - If $t = 1$, we obtain $7^x = 1 = 7^0$. Then $x = 0$.
    - If $t = 2$, then $7^x = 2$. Hence, $x = \log_7 2$.

    $\boxed{\text{Therefore, } x \in \{0, \log_7 2\}.}$

    Remark 2. Suppose that $a > 0, b > 0$ and $a \neq 1$. We obtain $a^k = b$ if and only if $k = \log_a b$.

Chapter 2. Exponential Functions

> **Example 22**
>
> Solve the following exponential equations:
>
> 1. $9^{x^2-1} - 36 \times 3^{x^2-3} + 3 = 0$
> 2. $3^{2x^2} - 2 \times 3^{x^2+x+6} + 3^{2(x+6)} = 0$
> 3. $4^{x^2+2} - 9 \times 2^{x^2+2} + 8 = 0$
> 4. $4^{x+\sqrt{x^2-2}} - 5 \times 2^{x-1+\sqrt{x^2-2}} = 6$

Solution. Solve the following exponential equations:

1. $9^{x^2-1} - 36 \times 3^{x^2-3} + 3 = 0$
   We have $9^{x^2-1} - 36 \times 3^{x^2-3} + 3 = 0$.
   Then
   $$3^{2(x^2-1)} - 36 \times 3^{-2} \times 3^{x^2-1} + 3 = 0$$
   or
   $$\left(3^{x^2-1}\right)^2 - 4 \times 3^{x^2-1} + 3 = 0.$$
   Let $t = 3^{x^2-1}$. Then $t > 0$ for all $x \in \mathbb{R}$. The given equation can be written as
   $$t^2 - 4t + 3 = 0.$$
   Hence, $(t-1)(t-3) = 0$. Then $\begin{bmatrix} t - 1 = 0 \\ t - 3 = 0 \end{bmatrix}$ or $\begin{bmatrix} t = 1 \\ t = 3 \end{bmatrix}$.

   - If $t = 1$, we obtain $3^{x^2-1} = 1 = 3^0$.
     Then $x^2 - 1 = 0$ or $x^2 = 1$.
     Therefore, $x \in \{-1, 1\}$.
   - If $t = 3$, we obtain $3^{x^2-1} = 3$.
     Then $x^2 - 1 = 1$ or $x^2 = 2$.
     Hence, $x \in \left\{-\sqrt{2}, \sqrt{2}\right\}$.

   > Therefore, $x \in \left\{-\sqrt{2}, -1, 1, \sqrt{2}\right\}$.

2. $3^{2x^2} - 2 \times 3^{x^2+x+6} + 3^{2(x+6)} = 0$
   The given equation can be written as
   $$\left(3^{x^2}\right)^2 - 2 \times 3^{x^2} \times 3^{x+6} + \left(3^{x+6}\right)^2 = 0.$$

2.3. How to Solve an Exponential Equation

Then $\left(3^{x^2} - 3^{x+6}\right)^2 = 0$. It follows that $3^{x^2} - 3^{x+6} = 0$.
Then $3^{x^2} = 3^{x+6}$.
We obtain $x^2 = x + 6$ or $x^2 - x - 6 = 0$.
It implies that $(x-3)(x+2) = 0$.
Consequently, $\begin{bmatrix} x+2=0 \\ x-3=0 \end{bmatrix}$ or $\begin{bmatrix} x=-2 \\ x=3 \end{bmatrix}$.

Therefore, $x \in \{-2, 3\}$.

3. $4^{x^2+2} - 9 \times 2^{x^2+2} + 8 = 0$
   We have $4^{x^2+2} - 9 \times 2^{x^2+2} + 8 = 0$. Then
   $$\left(2^{x^2+2}\right)^2 - 9 \times 2^{x^2+2} + 8 = 0.$$
   Let $t = 2^{x^2+2}$. We obtain $t \geq 2^2 = 4$ for all $x \in \mathbb{R}$. The given equation can be written as $t^2 - 9t + 8 = 0$ or $(t-1)(t-8) = 0$.
   Hence, $t - 8 = 0$ or $t = 8$ since $t \geq 4$.
   Then $2^{x^2+2} = 8$ or $2^{x^2+2} = 2^3$.
   It follows that $x^2 + 2 = 3$ or $x^2 = 1$.
   Therefore, $x \in \{-1, 1\}$.

4. $4^{x+\sqrt{x^2-2}} - 5 \times 2^{x-1+\sqrt{x^2-2}} = 6$
   The given equation is well-defined when $x^2 - 2 \geq 0$.
   We obtain $x^2 \geq 2$.
   It follows that $x \leq -\sqrt{2}$ or $x \geq \sqrt{2}$.
   We have
   $$4^{x+\sqrt{x^2-2}} - 5 \times 2^{x-1+\sqrt{x^2-2}} = 6$$
   or
   $$\left(2^{x+\sqrt{x^2-2}}\right)^2 - 5 \times 2^{-1} \times 2^{x+\sqrt{x^2-2}} - 6 = 0.$$
   Then $\left(2^{x+\sqrt{x^2-2}}\right)^2 - \frac{5}{2} \times 2^{x+\sqrt{x^2-2}} - 6 = 0$.
   It implies that $2\left(2^{x+\sqrt{x^2-2}}\right)^2 - 5 \times 2^{x+\sqrt{x^2-2}} - 12 = 0$.
   Let $t = 2^{x+\sqrt{x^2-2}}$. Then $t > 0$ for all $x \in \mathbb{R}$.
   We obtain $2t^2 - 5t - 12 = 0$ or $(t-4)(2t+3) = 0$.
   By knowing that $2t + 3 > 0$, it turns out that $t - 4 = 0$ or $t = 4$.
   For $t = 4$, we obtain $2^{x+\sqrt{x^2-2}} = 4 = 2^2$.
   Hence, $x + \sqrt{x^2 - 2} = 2$ or $\sqrt{x^2 - 2} = 2 - x$.

31

Chapter 2. Exponential Functions

- If $2-x < 0$ or $x > 2$, the given equation has no solutions.
- If $2-x \geq 0$ or $x \leq 2$, squaring both sides of the equation, we obtain

$$x^2 - 2 = (2-x)^2 \text{ or } x^2 - 2 = 4 - 4x + x^2.$$

Then $4x = 4 + 2 = 6$.

Therefore, $x = \dfrac{6}{4} = \dfrac{3}{2}$.

Example 23

Solve the following exponential equations:

1. $\left(\sqrt{\sqrt{7}-\sqrt{6}}\right)^x + \left(\sqrt{\sqrt{7}+\sqrt{6}}\right)^x = 2\sqrt{7}$

2. $\left(\sqrt{2-\sqrt{3}}\right)^x + 2\left(\sqrt{2+\sqrt{3}}\right)^x = 6$

Solution. Solve the following exponential equations:

1. $\left(\sqrt{\sqrt{7}-\sqrt{6}}\right)^x + \left(\sqrt{\sqrt{7}+\sqrt{6}}\right)^x = 2\sqrt{7}$

   Observe that

   $$\left(\sqrt{\sqrt{7}-\sqrt{6}}\right)^x \left(\sqrt{\sqrt{7}+\sqrt{6}}\right)^x$$
   $$= \left[\sqrt{\left(\sqrt{7}-\sqrt{6}\right)\left(\sqrt{7}+\sqrt{6}\right)}\right]^x$$
   $$= \left(\sqrt{\sqrt{7}^2 - \sqrt{6}^2}\right)^x$$
   $$= \left(\sqrt{7-6}\right)^x$$
   $$= \sqrt{1^x}$$
   $$= 1.$$

   Let $t = \left(\sqrt{\sqrt{7}-\sqrt{6}}\right)^x$. Then $t > 0$ for all $x \in \mathbb{R}$. It follows that $t\left(\sqrt{\sqrt{7}+\sqrt{6}}\right)^x = 1$. Hence, $\left(\sqrt{\sqrt{7}+\sqrt{6}}\right)^x = \dfrac{1}{t}$.

2.3. How to Solve an Exponential Equation

The given equation can be written as $t + \dfrac{1}{t} = 2\sqrt{7}$. We obtain

$$t^2 - 2\sqrt{7}t + 1 = 0.$$

The discriminant of the quadratic equation is

$$\begin{aligned}\Delta = b'^2 - ac \\ = \left(\sqrt{7}\right)^2 - (1)(1) \\ = 7 - 1 \\ = 6.\end{aligned}$$

The equation has two distinct real roots. They are

$$t_1 = \dfrac{-b' + \sqrt{\Delta'}}{a} = \dfrac{\sqrt{7} + \sqrt{6}}{1} = \sqrt{7} + \sqrt{6}$$

and

$$t_2 = \dfrac{-b' - \sqrt{\Delta'}}{a} = \dfrac{\sqrt{7} - \sqrt{6}}{1} = \sqrt{7} - \sqrt{6}.$$

- If $t = \sqrt{7} + \sqrt{6}$, we obtain

$$\begin{aligned}\left(\sqrt{\sqrt{7} - \sqrt{6}}\right)^x = \sqrt{7} + \sqrt{6} \\ = \dfrac{1}{\sqrt{7} - \sqrt{6}} \\ = \left(\sqrt{7} - \sqrt{6}\right)^{-1} \\ = \sqrt{\left(\sqrt{7} - \sqrt{6}\right)^{-2}}.\end{aligned}$$

Hence, $x = -2$.

- If $t = \sqrt{7} - \sqrt{6}$, it follows that

$$\left(\sqrt{\sqrt{7} - \sqrt{6}}\right)^x = \sqrt{7} - \sqrt{6} = \left(\sqrt{\sqrt{7} - \sqrt{6}}\right)^2.$$

Hence, $x = 2$.

$\boxed{\text{Therefore, } x \in \{-2, 2\}.}$

2. $\left(\sqrt{2-\sqrt{3}}\right)^x + 2\left(\sqrt{2+\sqrt{3}}\right)^x = 6+\sqrt{3}$

Observe that

$$\left(\sqrt{2-\sqrt{3}}\right)^x \left(\sqrt{2+\sqrt{3}}\right)^x$$
$$= \left[\sqrt{\left(2-\sqrt{3}\right)\left(2+\sqrt{3}\right)}\right]^x$$
$$= \left(\sqrt{2^2 - \sqrt{3}^2}\right)^x$$
$$= \left(\sqrt{4-3}\right)^x$$
$$= \sqrt{1^x}$$
$$= 1.$$

Let $t = \left(\sqrt{2-\sqrt{3}}\right)^x$. Then $t > 0$. It turns out that $t\left(\sqrt{2+\sqrt{3}}\right)^x = 1$. It follows that $t = \dfrac{1}{\left(\sqrt{2+\sqrt{3}}\right)^x}$. The given equation can be written as $t + \dfrac{2}{t} = 6+\sqrt{3}$. Then $t^2 - \left(6+\sqrt{3}\right)t + 2 = 0$. The discriminant of the quadratic equation is

$$\Delta = b^2 - 4ac$$
$$= \left(6+\sqrt{3}\right)^2 - 4\,(1)\,(2)$$
$$= 36 + 12\sqrt{3} + \sqrt{3}^2 - 8$$
$$= 31 + 12\sqrt{3}$$
$$= 27 + 12\sqrt{3} + 4$$
$$= \left(3\sqrt{3}\right)^2 + 2\left(3\sqrt{3}\right)(2) + 2^2$$
$$= \left(3\sqrt{3}+2\right)^2.$$

We obtain

$$t_1 = \frac{-b + \sqrt{\Delta}}{2a}$$

2.3. How to Solve an Exponential Equation

$$= \frac{6+\sqrt{3}+\sqrt{(3\sqrt{3}+2)^2}}{2}$$
$$= \frac{6+\sqrt{3}+3\sqrt{3}+2}{2}$$
$$= \frac{8+4\sqrt{3}}{2}$$
$$= \frac{2(4+2\sqrt{3})}{2}$$
$$= 4+2\sqrt{3}$$

and

$$t_2 = \frac{-b-\sqrt{\Delta}}{2a}$$
$$= \frac{6+\sqrt{3}-\sqrt{(3\sqrt{3}+2)^2}}{2}$$
$$= \frac{6+\sqrt{3}-3\sqrt{3}-2}{2}$$
$$= \frac{4-2\sqrt{3}}{2}$$
$$= \frac{2(2-\sqrt{3})}{2}$$
$$= 2-\sqrt{3}.$$

- If $t = 4+2\sqrt{3}$
$$= (2+\sqrt{3})^2$$
$$= \left(\frac{1}{2-\sqrt{3}}\right)^2$$
$$= (2-\sqrt{3})^{-2}$$

, we obtain $\left(\sqrt{2-\sqrt{3}}\right)^x = (2-\sqrt{3})^{-2} = \left(\sqrt{2-\sqrt{3}}\right)^{-4}$.

Hence, $x = -4$.

- If $t = 2-\sqrt{3} = \left(\sqrt{2-\sqrt{3}}\right)^2$, we obtain $\left(\sqrt{2-\sqrt{3}}\right)^x = \left(\sqrt{2-\sqrt{3}}\right)^2$. It follows that $x = 2$.

Therefore, $x \in \{-4, 2\}$.

Example 24

Solve the following exponential equations:

1. $\left(3 - \sqrt{5}\right)^x + \left(3 + \sqrt{5}\right)^x = 3 \times 2^x$
2. $\left(4 - \sqrt{7}\right)^x + \left(4 + \sqrt{7}\right)^x = 8 \times 3^{x-1}$

Solution. Solve the following exponential equations:

1. $\left(3 - \sqrt{5}\right)^x + \left(3 + \sqrt{5}\right)^x = 3 \times 2^x$
   We have
$$\left(3 - \sqrt{5}\right)^x + \left(3 + \sqrt{5}\right)^x = 3 \times 2^x$$
$$\frac{(3 - \sqrt{5})^x}{2^x} + \frac{(3 + \sqrt{5})^x}{2^x} = 3$$
$$\left(\frac{3 - \sqrt{5}}{2}\right)^x + \left(\frac{3 + \sqrt{5}}{2}\right)^x = 3.$$

Observe that
$$\left(\frac{3 - \sqrt{5}}{2}\right)^x \left(\frac{3 + \sqrt{5}}{2}\right)^x = \left[\left(\frac{3 - \sqrt{5}}{2}\right)\left(\frac{3 + \sqrt{5}}{2}\right)\right]^x$$
$$= \left(\frac{3^2 - \sqrt{5}^2}{4}\right)^x$$
$$= \left(\frac{9 - 5}{4}\right)^x$$
$$= 1^x$$
$$= 1.$$

Let $t = \left(\frac{3 - \sqrt{5}}{2}\right)^x$. Then $t > 0$ for all $x \in \mathbb{R}$.
It follows that $t\left(\frac{3 + \sqrt{5}}{2}\right)^x = 1$. Hence, $\left(\frac{3 + \sqrt{5}}{2}\right)^x = \frac{1}{t}$.

2.3. How to Solve an Exponential Equation

The given equation can be written as $t + \dfrac{1}{t} = 3$. We obtain $t^2 - 3t + 1 = 0$. The discriminant of the quadratic equation is

$$\Delta = b^2 - 4ac$$
$$= (-3)^2 - 4(1)(1)$$
$$= 9 - 4$$
$$= 5.$$

The equation has two distinct real roots. They are

$$t_1 = \frac{-b + \sqrt{\Delta}}{2a} = \frac{3 + \sqrt{5}}{2}$$

and

$$t_2 = \frac{-b - \sqrt{\Delta}}{2a} = \frac{3 - \sqrt{5}}{2}.$$

- If $t = \dfrac{3 + \sqrt{5}}{2}$, we obtain

$$\left(\frac{3 - \sqrt{5}}{2}\right)^x = \frac{3 + \sqrt{5}}{2} = \left(\frac{3 - \sqrt{5}}{2}\right)^{-1}.$$

Hence, $x = -1$.

- If $t = \dfrac{3 - \sqrt{5}}{2}$, we obtain $\left(\dfrac{3 - \sqrt{5}}{2}\right)^x = \dfrac{3 - \sqrt{5}}{2}.$

Hence, $x = 1$

$$\boxed{\text{Therefore, } x \in \{-1, 1\}.}$$

2. $\left(4 - \sqrt{7}\right)^x + \left(4 + \sqrt{7}\right)^x = 8 \times 3^{x-1}$

We have

$$\left(4 - \sqrt{7}\right)^x + \left(4 + \sqrt{7}\right)^x = 8 \times 3^{x-1}$$

$$\left(4 - \sqrt{7}\right)^x + \left(4 + \sqrt{7}\right)^x = \frac{8}{3} \times 3^x$$

$$\left(\frac{4 - \sqrt{7}}{3}\right)^x + \left(\frac{4 + \sqrt{7}}{3}\right)^x = \frac{8}{3}.$$

Observe that
$$\left(\frac{4-\sqrt{7}}{3}\right)^x \left(\frac{4+\sqrt{7}}{3}\right)^x = \left[\left(\frac{4-\sqrt{7}}{3}\right)\left(\frac{4+\sqrt{7}}{3}\right)\right]^x$$
$$= \left(\frac{4^2 - \sqrt{7}^2}{3^2}\right)^x$$
$$= \left(\frac{16-7}{9}\right)^x$$
$$= 1^x$$
$$= 1.$$

Let $t = \left(\frac{4-\sqrt{7}}{3}\right)^x$. Then $t > 0$ for all $x \in \mathbb{R}$. It follows that
$$\left(\frac{4+\sqrt{7}}{3}\right)^x = \frac{1}{t}.$$

The given equation can be written as $t + \frac{1}{t} = \frac{8}{3}$.
We obtain $3t^2 - 8t + 3 = 0$.
The discriminant of the quadratic equation is
$$\Delta' = b'^2 - ac$$
$$= (-4)^2 - (3)(3)$$
$$= 16 - 9$$
$$= 7.$$

We obtain
$$t_1 = \frac{-b' + \sqrt{\Delta'}}{a} = \frac{4+\sqrt{7}}{3}$$
and
$$t_2 = \frac{-b' - \sqrt{\Delta'}}{a} = \frac{4-\sqrt{7}}{3}.$$

- If $t = \frac{4+\sqrt{7}}{3}$, we obtain
$$\left(\frac{4-\sqrt{7}}{3}\right)^x = \frac{4+\sqrt{7}}{3}$$

2.4. Exponential Inequality

$$= \frac{1}{\left(\frac{4-\sqrt{7}}{3}\right)}$$

$$= \left(\frac{4-\sqrt{7}}{3}\right)^{-1}.$$

It implies that $x = -1$.

- If $t = \dfrac{4-\sqrt{7}}{3}$, we obtain $\left(\dfrac{4-\sqrt{7}}{3}\right)^x = \dfrac{4-\sqrt{7}}{3}$.

  Hence, $x = 1$.

Therefore, $x \in \{-1, 1\}$.

2.4 Exponential Inequality

Suppose that f is an exponential function which is defined by $f(x) = a^x$, where $a > 0$ and $a \neq 1$.

- If $a > 1$, we obtain f is an increasing function. In this case, $a^x > a^y$ if and only if $x > y$.

- If $0 < a < 1$, we obtain f is a decreasing function. In this case, $a^x > a^y$ if and only if $x < y$.

Hence, to solve an exponential inequality, we follow the following steps:

> **Theorem 3**
> Suppose that $a > 0$ and $a \neq 1$ such that $a^x > a^y$.
> - If $a > 1$, we obtain $x > y$.
> - If $0 < a < 1$, we obtain $x < y$.

The following examples will illustrate how to solve exponential inequalities:

> **Example 25**
> Solve the following exponential inequalities:

1. $3^x > 9$
2. $5^x < 25$
3. $\left(\dfrac{1}{2}\right)^x \leq \dfrac{1}{4}$
4. $\left(\dfrac{1}{3}\right)^x \geq \dfrac{1}{27}$
5. $\left(\dfrac{1}{5}\right)^x \geq 125$

Solution. Solve the following exponential inequalities:

1. $3^x > 9$
   We have $3^x > 9$. Then $3^x > 3^2$.
   Therefore, $x > 2$.

2. $5^x < 25$
   We have $5^x < 25$. Then $5^x < 5^2$.
   Therefore, $x < 2$.

3. $\left(\dfrac{1}{2}\right)^x \leq \dfrac{1}{4}$
   We have $\left(\dfrac{1}{2}\right)^x \leq \dfrac{1}{4}$. Then $\left(\dfrac{1}{2}\right)^x \leq \left(\dfrac{1}{2}\right)^2$.
   Therefore, $x \geq 2$.

4. $\left(\dfrac{1}{3}\right)^x \geq \dfrac{1}{27}$
   We have $\left(\dfrac{1}{3}\right)^x \geq \dfrac{1}{27}$. Then $\left(\dfrac{1}{3}\right)^x \geq \left(\dfrac{1}{3}\right)^3$.
   Therefore, $x \leq 3$.

5. $\left(\dfrac{1}{5}\right)^x \geq 125$
   We have $\left(\dfrac{1}{5}\right)^x \geq 125$.
   Then $\left(\dfrac{1}{5}\right)^x \geq 5^3$. It follows that $\left(\dfrac{1}{5}\right)^x \geq \left(\dfrac{1}{5}\right)^{-3}$.
   Therefore, $x \leq -3$.

2.4. Exponential Inequality

Example 26

Solve the following exponential inequalities:

1. $\sqrt{3^x} \geq \sqrt[3]{9}$
2. $3^{x^2-2x} < 1$
3. $7^{\frac{x-3}{x}} > \frac{1}{7}$
4. $2^{x^2-x} \geq 4$
5. $8^{x^2+1} - 64^x \geq 0$

Solution. Solve the following exponential inequalities:

1. $\sqrt{3^x} \geq \sqrt[3]{9}$
   We have $\sqrt{3^x} \geq \sqrt[3]{9}$. Then $3^{\frac{x}{2}} \geq 9^{\frac{1}{3}}$. It follows that $3^{\frac{x}{2}} \geq 3^{\frac{2}{3}}$.
   It implies that $\frac{x}{2} \geq \frac{2}{3}$. We obtain $x \geq \frac{4}{3}$.

   Therefore, $x \geq \frac{4}{3}$.

2. $3^{x^2-2x} < 1$
   We have $3^{x^2-2x} < 1$. Then $3^{x^2-2x} < 3^0$.
   It follows that $x^2 - 2x < 0$.
   If $x^2 - 2x = 0$, we obtain $x(x-2) = 0$. Hence, $\begin{bmatrix} x = 0 \\ x - 2 = 0 \end{bmatrix}$.

   It turns out that $\begin{bmatrix} x = 0 \\ x = 2 \end{bmatrix}$.

   Sign table of $x^2 - 2x$:

x	$-\infty$		0		2		$+\infty$
$x^2 - 2x$		+	0	−	0	+	

   Therefore, $x \in (0, 2)$.

3. $7^{\frac{x-3}{x}} > \frac{1}{7}$
   We have $7^{\frac{x-3}{x}} > \frac{1}{7}$. Then $7^{\frac{x-3}{x}} > 7^{-1}$. It follows that
   $$\frac{x-3}{x} > -1.$$

We obtain $\dfrac{x-3}{x} + 1 > 0$.

Hence, $\dfrac{2x-3}{x} > 0$.

If $2x - 3 = 0$, we obtain $x = \dfrac{3}{2}$.

Sign table of $\dfrac{2x-3}{x}$:

x	$-\infty$		0		$\dfrac{3}{2}$		$+\infty$
$2x-3$		$-$		$-$	0	$+$	
x		$-$	0	$+$		$+$	
$\dfrac{2x-3}{x}$		$+$	$\|$	$-$	0	$+$	

Therefore, $x \in (-\infty, 0) \cup \left(\dfrac{3}{2}, +\infty\right)$.

4. $2^{x^2-x} \geq 4$

We have $2^{x^2-x} \geq 4$. Then $2^{x^2-x} \geq 2^2$.

It follows that $x^2 - x \geq 2$. Hence, $x^2 - x - 2 \geq 0$.

If $x^2 - x - 2 = 0$, we obtain $(x+1)(x-2) = 0$. Then

$$\left[\begin{array}{l} x+1=0 \\ x-2=0 \end{array}\right. \text{ or } \left[\begin{array}{l} x=-1 \\ x=2 \end{array}\right..$$

x	$-\infty$		-1		2		$+\infty$
x^2-x-2		$+$	0	$-$	0	$+$	

Therefore, $x \in (-\infty, -1] \cup [2, +\infty)$.

5. $8^{x^2+1} - 64^x \geq 0$

We have $8^{x^2+1} - 64^x \geq 0$. Then $8^{x^2+1} \geq 64^x$.

It follows that $8^{x^2+1} \geq 8^{2x}$.

2.4. Exponential Inequality

We obtain $x^2 + 1 \geq 2x$. Then $x^2 - 2x + 1 \geq 0$.
Hence, $(x-1)^2 \geq 0$. The last inequality holds for all $x \in \mathbb{R}$.
Therefore, $x \in \mathbb{R}$.

> **Example 27**
> Solve the following exponential inequalities:
>
> 1. $16^x > 0.125$
> 2. $2^x > -8$
> 3. $27^x \times 3^{1-x} < \dfrac{1}{3}$
> 4. $0.2^x \leq 25$
> 5. $3^x \leq \sqrt[3]{9}$
> 6. $\left(\dfrac{1}{2}\right)^x > \sqrt[3]{\dfrac{1}{4}}$

Solution. Solve the following exponential inequalities:

1. $16^x > 0.125$
   We have $16^x > 0.125$. Then $16^x > \dfrac{125}{1000}$.
   It follows that $4^{2x} > \dfrac{1}{4}$.
   We obtain $4^{2x} > 4^{-1}$.
   It implies that $2x > -1$. Then $x > -\dfrac{1}{2}$.
   Therefore, $x > -\dfrac{1}{2}$.

2. $2^x > -8$
   By knowing that $2^x > 0$ for all $x \in \mathbb{R}$, we obtain $2^x > -8$ for all $x \in \mathbb{R}$.
   Therefore, $x \in \mathbb{R}$.

3. $27^x \times 3^{1-x} < \dfrac{1}{3}$
   We have $27^x \times 3^{1-x} < \dfrac{1}{3}$. It follows that $3^{3x} \times 3^{1-x} < 3^{-1}$.
   Then $3^{3x+1-x} < 3^{-1}$.
   We obtain $3^{2x+1} < 3^{-1}$.
   Then $2x + 1 < -1$. It turns out that $2x < -2$.
   Therefore, $x < -1$.

4. $0.2^x \leq 25$

   We have $0.2^x \leq 25$. Then $\left(\dfrac{2}{10}\right)^x \leq 25$.

   It follows that $\left(\dfrac{1}{5}\right)^x \leq 5^2$.

   We obtain $\left(\dfrac{1}{5}\right)^x \leq \left(\dfrac{1}{5}\right)^{-2}$.

   Therefore, $x \geq -2$.

5. $3^x \leq \sqrt[3]{9}$

   We have $3^x \leq \sqrt[3]{9}$. Then $3^x \leq \sqrt[3]{3^2}$. It implies that $3^x \leq 3^{\frac{2}{3}}$.

   Therefore, $x \leq \dfrac{2}{3}$.

6. $\left(\dfrac{1}{2}\right)^x > \sqrt[3]{\dfrac{1}{4}}$

   We have $\left(\dfrac{1}{2}\right)^x > \sqrt[3]{\dfrac{1}{4}}$. Then $\left(\dfrac{1}{2}\right)^x > \sqrt[3]{\left(\dfrac{1}{2}\right)^2}$.

   It follows that $\left(\dfrac{1}{2}\right)^x > \left(\dfrac{1}{2}\right)^{\frac{2}{3}}$.

   Therefore, $x < \dfrac{2}{3}$.

Example 28

Solve the following exponential inequalities:

1. $\left(\dfrac{3}{4}\right)^{6x+10-x^2} < \dfrac{27}{64}$

2. $\left(\dfrac{1}{3}\right)^{x^2+2x} < \left(\dfrac{1}{9}\right)^{16-x}$

3. $\dfrac{6}{3^x - 1} < 3^x$

Solution. Solve the following exponential inequalities:

1. $\left(\dfrac{3}{4}\right)^{6x+10-x^2} < \dfrac{27}{64}$

2.4. Exponential Inequality

We have $\left(\dfrac{3}{4}\right)^{6x+10-x^2} < \dfrac{27}{64}$. Then $\left(\dfrac{3}{4}\right)^{6x+10-x^2} < \left(\dfrac{3}{4}\right)^3$.

We obtain $6x + 10 - x^2 > 3$. It follows that $x^2 - 6x - 7 < 0$.
If $x^2 - 6x - 7 = 0$, we obtain $(x+1)(x-7) = 0$.
Then $\begin{bmatrix} x+1=0 \\ x-7=0 \end{bmatrix}$. Hence, $\begin{bmatrix} x=-1 \\ x=7 \end{bmatrix}$.

Sign table of $x^2 - 6x - 7$:

x	$-\infty$		-1		7		$+\infty$
$x^2 - 6x - 7$		$+$	0	$-$	0	$+$	

Therefore, $x \in (-1, 7)$.

2. $\left(\dfrac{1}{3}\right)^{x^2+2x} < \left(\dfrac{1}{9}\right)^{16-x}$

We have $\left(\dfrac{1}{3}\right)^{x^2+2x} < \left(\dfrac{1}{9}\right)^{16-x}$.

Then $\left(\dfrac{1}{3}\right)^{x^2+2x} < \left(\dfrac{1}{3}\right)^{2(16-x)}$.

It follows that

$$\left(\dfrac{1}{3}\right)^{x^2+2x} < \left(\dfrac{1}{3}\right)^{32-2x}.$$

We obtain $x^2 + 2x > 32 - 2x$. Then $x^2 + 4x - 32 > 0$.
If $x^2 + 4x - 32 = 0$, we obtain $(x+8)(x-4) = 0$. Then $\begin{bmatrix} x+8=0 \\ x-4=0 \end{bmatrix}$. It implies that $\begin{bmatrix} x=-8 \\ x=4 \end{bmatrix}$.

Sign table of $x^2 + 4x - 32$:

x	$-\infty$		-8		4		$+\infty$
$x^2 + 4x - 32$		$+$	0	$-$	0	$+$	

Therefore, $x \in (-\infty, -8) \cup (4, +\infty)$.

3. $\dfrac{6}{3^x - 1} < 3^x$

We have $\dfrac{6}{3^x - 1} < 3^x$. The given inequalities is well-defined if and only if $3^x - 1 \neq 0$. Then $3^x \neq 1 = 3^0$.
It follows that $x \neq 0$.
Let $t = 3^x$. It follows that $t > 0$ for all $x \in \mathbb{R}$.
The given inequality can be written as $\dfrac{6}{t-1} < t$.
We obtain $t - \dfrac{6}{t-1} > 0$.
Then $\dfrac{t^2 - t - 6}{t-1} > 0$. We obtain $\dfrac{(t-3)(t+2)}{t-1} > 0$.
By knowing that $t+2 > 0$ for all $x \in \mathbb{R}$, we obtain $\dfrac{t-3}{t-1} > 0$.

- If $t - 3 = 0$, we obtain $t = 3$.
- If $t - 1 = 0$, we obtain $t = 1$.

Sign table of $\dfrac{t-3}{t-1}$:

t	$-\infty$		1		3		$+\infty$
$t - 3$		$-$		$-$	0	$+$	
$t - 1$		$-$	0	$+$		$+$	
$\dfrac{t-3}{t-1}$		$+$		$-$	0	$+$	

It turns out that $t \in (-\infty, 1) \cup (3, +\infty)$. We obtain $t < 1$ or $t > 3$.

- If $t < 1$, we obtain $3^x < 1 = 3^0$. Then $x < 0$.
- If $t > 3$, we obtain $3^x > 3$. Hence, $x > 1$.

$\boxed{\text{Therefore, } x \in (-\infty, 0) \cup (1, +\infty).}$

Chapter 3

Logarithmic Functions

In the previous chapter, we learned about exponential functions and their properties. In this chapter, we will introduce readers a new kind of functions which is called logarithmic functions.

3.1 Definition

> **Definition 4**
> Suppose that a, b and x are real numbers such that $a > 0$ and $a^x = b$. In this case, we say that x is the logarithm of b to the base a. It is denoted by $x = \log_a b$.

> **Example 29**
> Compute the following expressions:
>
> 1. $\log_2 1$
> 2. $\log_2 2$
> 3. $\log_3 9$
> 4. $\log_3 \frac{1}{27}$
> 5. $\log_4 \frac{1}{64}$

Solution. Compute the following expressions:

1. $\log_2 1$
   Let $x = \log_2 1$. By definition, we have $2^x = 1$. It follows that $2^x = 2^0$. Then $x = 0$.
   Therefore, $\log_2 1 = 0.$

2. $\log_2 2$

   Let $x = \log_2 2$. By definition, we have $2^x = 2$. Then $x = 1$.
   $\boxed{\text{Therefore, } \log_2 2 = 1.}$

3. $\log_3 9$

   Let $x = \log_3 9$. By definition, we have $3^x = 9 = 3^2$. Then $x = 2$. $\boxed{\text{Therefore, } \log_3 9 = 2.}$

4. $\log_3 \dfrac{1}{27}$

   Let $x = \log_3 \dfrac{1}{27}$. By definition, we have $3^x = \dfrac{1}{27} = 3^{-3}$. Then $x = -3$.
   $\boxed{\text{Therefore, } \log_3 \dfrac{1}{27} = -3.}$

5. $\log_4 \dfrac{1}{64}$

   Let $x = \log_4 \dfrac{1}{64}$. By definition, we have $4^x = \dfrac{1}{64} = 4^{-3}$. Then $x = -3$.
   $\boxed{\text{Therefore, } \log_4 \dfrac{1}{64} = -3.}$

3.2 Logarithmic Functions

> **Definition 5**
>
> Let $a > 0$ and $a \neq 1$. The logarithmic function to the base a is a function f which is defined by
> $$f : (0, +\infty) \to \mathbb{R}$$
> $$x \mapsto f(x) = \log_a x.$$

Example 30

Find the domain of the following functions:

1. $f(x) = \log_2(x-1)$
2. $f(x) = \log_3(1+x)$
3. $f(x) = \log_4(3x-2)$
4. $f(x) = \log_5(4x-1)$

3.2. Logarithmic Functions

Solution. Find the domain of the following functions:

1. $f(x) = \log_2(x-1)$
   f is well-defined if and only if $x - 1 > 0$. Then $x > 1$.
   $\boxed{\text{Therefore, } \mathbb{D}_f = (1, +\infty).}$

2. $f(x) = \log_3(1+x)$
   f is well-defined if and only if $1 + x > 0$. Then $x > -1$.
   $\boxed{\text{Therefore, } \mathbb{D}_f = (-1, +\infty).}$

3. $f(x) = \log_4(3x-2)$
   f is well-defined if and only if $3x - 2 > 0$. Then $x > \dfrac{2}{3}$.
   $\boxed{\text{Therefore, } \mathbb{D}_f = \left(\dfrac{2}{3}, +\infty\right).}$

4. $f(x) = \log_5(4x-1)$
   f is well-defined if and only if $4x - 1 > 0$. Then $x > \dfrac{1}{4}$.
   $\boxed{\text{Therefore, } \mathbb{D}_f = \left(\dfrac{1}{4}, +\infty\right).}$

> **Example 31**
>
> Below are some examples of logarithmic functions:
>
> - $f(x) = \log_2 x$
> - $f(x) = \log_3 x$
> - $f(x) = \log_5 x$
> - $f(x) = \log_7 x$

3.2.1 Propertiies of Logarithm

> **General 5**
>
> For all positive real numbers x, y and a, where $a \neq 1$, we obtain
>
> 1. $\log_a 1 = 0$
> 2. $\log_a a = 1$
> 3. $\log_a(xy) = \log_a x + \log_a y$

4. $\log_a \left(\dfrac{x}{y}\right) = \log_a x - \log_a y$

5. $\log_a x^n = n \log_a x$

6. $\log_a a^n = n$

7. $\log_a x = \dfrac{1}{\log_x a}$

8. $a^{\log_a x} = x$

9. $\log_{a^\alpha} b^\beta = \dfrac{\beta}{\alpha} \log_a b$, where $\alpha \neq 0$

Proof. Prove that

1. $\log_a 1 = 0$
   Let $y = \log_a 1$. By definition, we have $a^y = 1 = a^0$. Then $y = 0$.
   $\boxed{\text{Therefore, } \log_a 1 = 0.}$

2. $\log_a a = 1$
   Let $y = \log_a a$. Then $a^y = a$. It follows that $y = 1$.
   $\boxed{\text{Therefore, } \log_a a = 1.}$

3. $\log_a (xy) = \log_a x + \log_a y$
   Let $p = \log_a x$ and $q = \log_a y$. We obtain $a^p = x$ and $a^q = y$. Then $xy = a^{p+q}$. By definition, we have
   $$\log_a (xy) = p + q = \log_a x + \log_a y.$$
   $\boxed{\text{Therefore, } \log_a (xy) = \log_a x + \log_a y.}$

4. $\log_a \left(\dfrac{x}{y}\right) = \log_a x - \log_a y$
   Using the notation in 3, we have $\dfrac{x}{y} = \dfrac{a^p}{a^q} = a^{p-q}$.
   $$\log_a \left(\dfrac{x}{y}\right) = p - q = \log_a x - \log_a y.$$

3.2. Logarithmic Functions

$$\boxed{\text{Therefore, } \log_a\left(\frac{x}{y}\right) = \log_a x - \log_a y.}$$

5. $\log_a x^n = n\log_a x$
   From property 3, we have

$$\log_a x^n = \log_a \underbrace{(x \times x \times ... \times x)}_{n}$$
$$= \underbrace{\log_a x + \log_a x + ... + \log_a x}_{n}$$
$$= n\log_a x.$$

$$\boxed{\text{Therefore, } \log_a x^n = n\log_a x.}$$

6. $\log_a a^n = n$
   From property 5, we have $\log_a a^n = n\log_a a$. By knowing that $\log_a a = 1$, we obtain $\log_a a^n = n$.

7. $\log_a x = \dfrac{1}{\log_x a}$
   Let $t = \log_a x$. By definition, we have $a^t = x$. Then $x^{\frac{1}{t}} = a$.
   It follows that $\log_x a = \dfrac{1}{t} = \dfrac{1}{\log_a x}$.

$$\boxed{\text{Therefore, } \log_a x = \dfrac{1}{\log_x a}.}$$

8. $a^{\log_a x} = x$
   Let $t = \log_a x$. Then $a^t = x$. It follows that $a^{\log_a x} = x$.

9. $\log_{a^\alpha} b^\beta = \dfrac{\beta}{\alpha}\log_a b$
   If $b = 1$, the given identity holds. If $b \neq 1$, from property 5 and 7, we obtain

$$\log_{a^\alpha} b^\beta = \beta \log_{a^\alpha} b$$
$$= \beta \frac{1}{\log_b a^\alpha}$$
$$= \frac{\beta}{\alpha} \frac{1}{\log_b a}$$

$$= \frac{\beta}{\alpha} \log_a b.$$

Therefore, $\log_{a^\alpha} b^\beta = \dfrac{\beta}{\alpha} \log_a b.$

□

Example 32

Given a function f which is defined by $f(x) = \log_a x$, where $a > 0, a \neq 1$ and $x > 0$. Prove that

1. $f(xy) = f(x) + f(y)$
2. $f\left(\dfrac{x}{y}\right) = f(x) - f(y)$

Solution. Prove that

1. $f(xy) = f(x) + f(y)$
   We have
   $$\begin{aligned} f(xy) &= \log_a xy \\ &= \log_a x + \log_a y \\ &= f(x) + f(y). \end{aligned}$$

   Therefore, $f(xy) = f(x) + f(y).$

2. $f\left(\dfrac{x}{y}\right) = f(x) - f(y)$
   We have
   $$\begin{aligned} f\left(\dfrac{x}{y}\right) &= \log_a \dfrac{x}{y} \\ &= \log_a x - \log_a y \\ &= f(x) - f(y). \end{aligned}$$

   Therefore, $f\left(\dfrac{x}{y}\right) = f(x) - f(y).$

3.2. Logarithmic Functions

> **General 6**
>
> **(Changing Base Formula)**
> For all positive real numbers a, b and c, where $a \neq 1$ and $c \neq 1$, we obtain $\log_a b = \dfrac{\log_c b}{\log_c a}$.

Proof. Let $t = \log_a b$. By definition, we have $a^t = b$. $\log_c a^t = \log_c b$. Using logarithmic property, we obtain $t \log_c a = \log_c b$. Then $t = \dfrac{\log_c b}{\log_c a}$.

Therefore, $\log_a b = \dfrac{\log_c b}{\log_c a}$. □

> **Example 33**
>
> Compute the following expressions:
>
> 1. $\log_3 9$
> 2. $\log_2 16$
> 3. $\log_4 0.25$
> 4. $\log_2 \dfrac{1}{32}$
> 5. $\log_3 \dfrac{1}{81}$

Solution. 1. $\log_3 9$
We have $\log_3 9 = \log_3 3^2 = 2$.
Therefore, $\log_3 9 = 2$.

2. $\log_2 16$
We have $\log_2 16 = \log_2 2^4 = 4$.
Therefore, $\log_2 16 = 4$.

3. $\log_4 0.25$
We have $\log_4 0.25 = \log_4 \dfrac{1}{4} = \log_4 4^{-1} = -1$.
Therefore, $\log_4 0.25 = -1$.

4. $\log_2 \dfrac{1}{32}$
We have $\log_2 \dfrac{1}{32} = \log_2 \dfrac{1}{2^5} = \log_2 2^{-5} = -5$.

> Therefore, $\log_2 \dfrac{1}{32} = -5$.

5. $\log_3 \dfrac{1}{81}$

   We have $\log_3 \dfrac{1}{81} = \log_3 \dfrac{1}{3^4} = \log_3 3^{-4} = -4$.

   > Therefore, $\log_3 \dfrac{1}{81} = -4$.

Example 34

Compute the following expressions:

1. $\log_3 6 - \log_3 2$
2. $\log_5 15 - \log_5 3$
3. $\log_7 14 - \log_7 2$
4. $\log_3 12 - 2\log_3 2$
5. $2\log_6 9 - \log_6 27 + \log_6 2$

Solution. Compute the following expressions:

1. $\log_3 6 - \log_3 2$

   We have $\log_3 6 - \log_3 2 = \log_3 \dfrac{6}{2} = \log_3 3 = 1$.

   > Therefore, $\log_3 6 - \log_3 2 = 1$.

2. $\log_5 15 - \log_5 3$

   We have $\log_5 15 - \log_5 3 = \log_5 \dfrac{15}{3} = \log_5 5 = 1$.

   > Therefore, $\log_5 15 - \log_5 3 = 1$.

3. $\log_7 14 - \log_7 2$

   We have $\log_7 14 - \log_7 2 = \log_7 \dfrac{14}{2} = \log_7 7 = 1$.

   > Therefore, $\log_7 14 - \log_7 2 = 1$.

4. $\log_3 12 - 2\log_3 2$

   We have

   $$\begin{aligned}\log_3 12 - 2\log_3 2 &= \log_3 12 - \log_3 2^2 \\ &= \log_3 12 - \log_3 4 \\ &= \log_3 \dfrac{12}{4}\end{aligned}$$

3.2. Logarithmic Functions

$$= \log_3 3$$
$$= 1.$$

Therefore, $\log_3 12 - 2\log_3 2 = 1.$

5. $2\log_6 9 - \log_6 27 + \log_6 2$
   We have

$$2\log_6 9 - \log_6 27 + \log_6 2$$
$$= \log_6 9^2 - \log_6 27 + \log_6 2$$
$$= \log_6 \left(\frac{81 \times 2}{27}\right)$$
$$= \log_6 6$$
$$= 1.$$

Therefore, $2\log_6 9 - \log_6 27 + \log_6 2 = 1.$

Example 35

Compute $\log_2^2 7 - \log_2 \frac{7}{2} \times \log_2 14.$

Solution. We have

$$\log_2^2 7 - \log_2 \frac{7}{2} \times \log_2 14$$
$$= \log_2^2 7 - (\log_2 7 - \log_2 2)(\log_2 7 + \log_2 2)$$
$$= \log_2^2 7 - (\log_2 7 - 1)(\log_2 7 + 1)$$
$$= \log_2^2 7 - (\log_2^2 7 - 1)$$
$$= \log_2^2 7 - \log_2^2 7 + 1$$
$$= 1.$$

Therefore, $\log_2^2 7 - \log_2 \frac{7}{2} \times \log_2 14 = 1.$

> **Example 36**
> Compute $\log_2 \dfrac{1}{2} + \log_2 \dfrac{2}{3} + \ldots + \log_2 \dfrac{2021}{2022}$.

Solution. We have

$$\log_2 \frac{1}{2} + \log_2 \frac{2}{3} + \ldots + \log_2 \frac{2021}{2022}$$
$$= \log_2 \left(\frac{1}{2} \times \frac{2}{3} \times \ldots \times \frac{2021}{2022} \right)$$
$$= \log_2 \frac{1}{2022}.$$

Therefore, $\log_2 \dfrac{1}{2} + \log_2 \dfrac{2}{3} + \ldots + \log_2 \dfrac{2021}{2022} = \log_2 \dfrac{1}{2022}$.

> **Example 37**
> Compute $\log_5 \dfrac{20}{9} + \log_5 \dfrac{6}{25} + \log_5 \dfrac{75}{8}$.

Solution. We have

$$\log_5 \frac{20}{9} + \log_5 \frac{6}{25} + \log_5 \frac{75}{8}$$
$$= \log_5 \left(\frac{20}{9} \times \frac{6}{25} \times \frac{75}{8} \right)$$
$$= \log_5 \left(\frac{2^2 \times 5 \times 2 \times 3 \times 5^2 \times 3}{3^2 \times 5^2 \times 2^3} \right)$$
$$= \log_5 \left(\frac{2^3 \times 3^2 \times 5^3}{2^3 \times 3^2 \times 5^2} \right)$$
$$= \log_5 5$$
$$= 1.$$

Therefore, $\log_5 \dfrac{20}{9} + \log_5 \dfrac{6}{25} + \log_5 \dfrac{75}{8} = 1$.

3.2. Logarithmic Functions

Example 38

Compute $\log_2\log_2\log_3\log_2 512$.

Solution. We have

$$\begin{aligned}\log_2\log_2\log_3\log_2 512 &= \log_2\log_2\log_3\log_2 2^9 \\ &= \log_2\log_2\log_3 9 \\ &= \log_2\log_2\log_3 3^2 \\ &= \log_2\log_2 2 \\ &= \log_2 1 \\ &= 0.\end{aligned}$$

Therefore, $\log_2\log_2\log_3\log_2 512 = 0$.

Example 39

Compute $3^{2\log_3 2}$.

Solution. We have $3^{2\log_3 2} = \left(3^{\log_3 2}\right)^2 = 2^2 = 4$.

Therefore, $3^{2\log_3 2} = 4$.

Example 40

Compute the following expressions:

1. $\log_2 \dfrac{1}{32}$

2. $\log 0.00001$

3. $\log_{\sqrt{3}-\sqrt{2}}\left(\sqrt{3}+\sqrt{2}\right)$

4. $\log_9 0.\overline{7}$

5. $\log_{\sqrt{2}} 2 + \log_{\sqrt{3}} 9$

Solution. Compute the following expressions:

1. $\log_2 \dfrac{1}{32}$

   We have $\log_2 \dfrac{1}{32} = \log_2 \dfrac{1}{2^5} = \log_2 2^{-5} = -5$.

   Therefore, $\log_2 \dfrac{1}{32} = -5$.

57

Chapter 3. Logarithmic Functions

2. $\log 0.00001$
   We have $\log 0.00001 = \log 10^{-5} = -5$
   Therefore, $\log 0.00001 = -5.$

3. $\log_{\sqrt{3}-\sqrt{2}}\left(\sqrt{3}+\sqrt{2}\right)$
   We have
   $$\frac{1}{\sqrt{3}+\sqrt{2}} = \frac{\sqrt{3}-\sqrt{2}}{\left(\sqrt{3}+\sqrt{2}\right)\left(\sqrt{3}-\sqrt{2}\right)}$$
   $$= \frac{\sqrt{3}-\sqrt{2}}{\sqrt{3^2}-\sqrt{2^2}}$$
   $$= \frac{\sqrt{3}-\sqrt{2}}{3-2}$$
   $$= \sqrt{3}-\sqrt{2}.$$
   Then $\sqrt{3}+\sqrt{2} = \dfrac{1}{\sqrt{3}-\sqrt{2}} = \left(\sqrt{3}-\sqrt{2}\right)^{-1}.$
   It follows that
   $$\log_{\sqrt{3}-\sqrt{2}}\left(\sqrt{3}+\sqrt{2}\right) = \log_{\sqrt{3}-\sqrt{2}}\left(\sqrt{3}-\sqrt{2}\right)^{-1} = -1.$$

   Therefore, $\log_{\sqrt{3}-\sqrt{2}}\left(\sqrt{3}+\sqrt{2}\right) = -1.$

4. $\log_{\frac{9}{7}} 0.\overline{7}$
   We have $0.\overline{7} = 7 \times 0.\overline{1} = 7 \times \dfrac{1}{9} = \dfrac{7}{9}.$
   It implies that $\log_{\frac{9}{7}} 0.\overline{7} = \log_{\frac{9}{7}} \dfrac{7}{9} = \log_{\frac{9}{7}} \left(\dfrac{9}{7}\right)^{-1} = -1.$
   Therefore, $\log_{\frac{9}{7}} 0.\overline{7} = -1.$

5. $\log_{\sqrt{2}} 2 + \log_{\sqrt{3}} 9$
   We have
   $$\log_{\sqrt{2}} 2 + \log_{\sqrt{3}} 9 = \log_{2^{\frac{1}{2}}} 2 + \log_{3^{\frac{1}{2}}} 3^2$$
   $$= \frac{1}{\frac{1}{2}} \log_2 2 + \frac{2}{\frac{1}{2}} \log_3 3$$
   $$= 2 + 2(2)$$

58

3.2. Logarithmic Functions

$$= 6.$$

Therefore, $\log_{\sqrt{2}} 2 + \log_{\sqrt{3}} 9 = 6$.

Example 41

Prove that $\log_3 2$ is an irrational number.

Solution. Suppose that $\log_3 2$ is a rational number. Then there exist two positive real numbers p and q such that $q \neq 0$ and $\log_3 2 = \frac{p}{q}$. Then $3^{\frac{p}{q}} = 2$. It follows that $3^p = 2^q$, a contradiction to the fact that 3^p is an odd number and 2^q is an even number.
Therefore, $\log_3 2$ is an irrational number.

Example 42

Given $\log_2 a = x$ and $\log_5 b = x$. Find 100^x in terms of a and b.

Solution. Find 100^x in terms of a and b.
We have $\log_2 a = x$ and $\log_5 b = x$. Then $2^x = a$ and $5^x = b$. Hence,

$$100^x = \left(2^2 \times 5^2\right)^x = (2^x)^2 \times (5^x)^2 = a^2 b^2.$$

Therefore, $100^x = a^2 b^2$.

3.2.2 Graphs of Logarithmic Functions

Example 43

Graph the logarithmic function which is defined by $y = \log_2 x$.

Solution. To draw the graph of f, we first construct the following table of values for it:

x	1/2	1	2	4
y	-1	0	1	2

Chapter 3. Logarithmic Functions

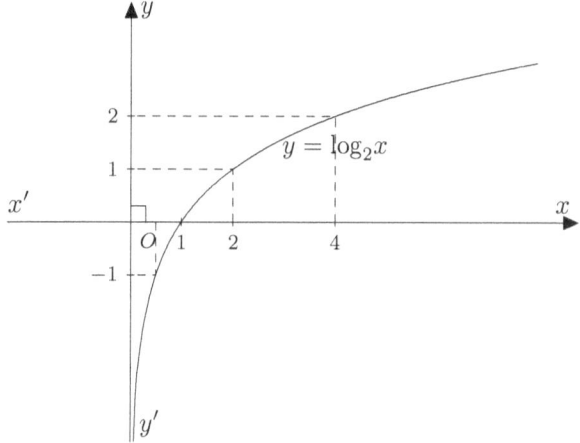

We see that the graph of $y = \log_2 x$ is increasing from left to right. Hence, y is an increasing function.

In general, the graph C of a function f which is defined by $y = \log_a x$, where $a > 1$ has the following properties:

1. The graph C always passes through $(1, 0)$.
2. The graph C has outputs for all $x > 0$.
3. The outputs of f are on $(-\infty, +\infty)$.
4. The graph C is increasing from left to right.
5. The graph C approaches the y axis when x tend to 0 from the right.
6. The graph C tends to $+\infty$ when x approaches $+\infty$.

> **Example 44**
>
> Graph the logarithmic function which is defined by $y = \log_{\frac{1}{2}} x$.

Solution. To draw the graph of f, we first construct the following table of values for it:

x	1/2	1	2	4
y	1	0	−1	−2

3.2. Logarithmic Functions

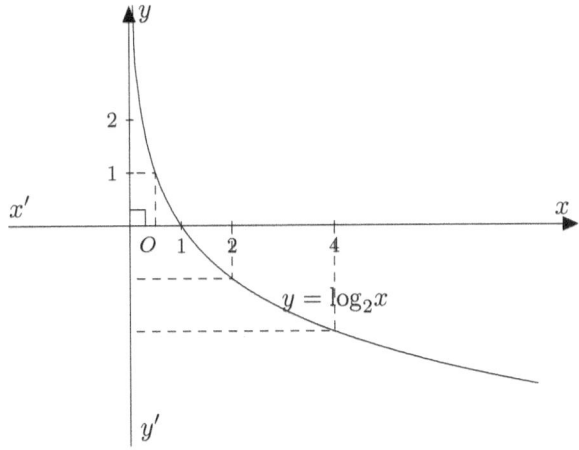

We see that the graph of $y = \log_{\frac{1}{2}} x$ is decreasing from left to right. Hence, y is a decreasing function. Generally, the graph C of a function f which is defined by $y = \log_a x$, where $0 < a < 1$ has the following properties:

1. The graph C always passes through $(1, 0)$.

2. The graph C has outputs for all $x > 0$.

3. The outputs of f are on $(-\infty, +\infty)$.

4. The graph C is decreasing from left to right.

5. The graph C approaches the y axis when x tend to 0 from the right.

6. The graph C tends to the $-\infty$ when x approaches $+\infty$.

General 7

Given a logarithmic function f which is defined by

$$f : (0, +\infty) \to \mathbb{R}$$
$$x \mapsto f(x) = \log_a x \text{ , where } a > 0 \text{ and } a \neq 1.$$

- If $a > 1$, then f is an increasing function.

- If $0 < a < 1$, then f is a decreasing function.

3.3 Logarithmic Equations and Inequalities:

3.3.1 Logarithmic Equations

3.3.2 Existence of Roots of Logarithmic Equations

A logarithmic equation is an equation that can be written in the form of $\log_a x = b$, where $a > 0$ and $a \neq 1$. The number of solutions to the equation $\log_a x = b$ is the number of the intersection points of the graphs of the function $y = \log_a x$ and the horizontal line $y = b$. There are two cases to consider:
If $a > 1$, then y is an increasing function. In this case, we obtain the following graph:

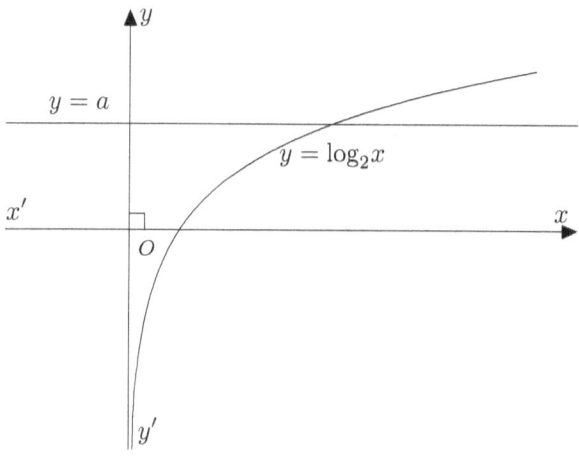

If $0 < a < 1$, then $f(x) = a^x$ is a decreasing function. In this case, we obtain the following graph:

3.3. Logarithmic Equations and Inequalities:

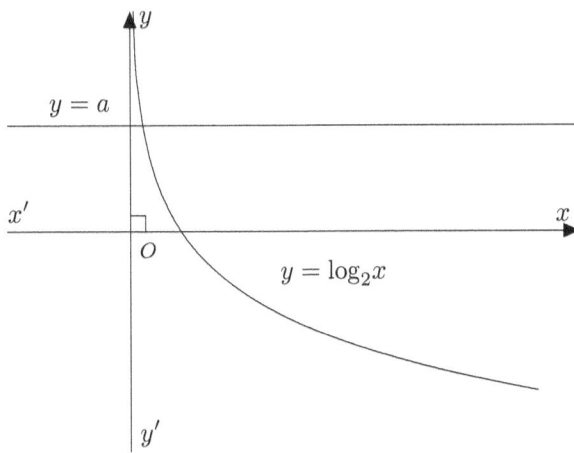

From the above graphs, we see that the graph of the function $y = \log_a x$ and the horizontal line $y = b$ always cut at a point. Therefore, the logarithmic equation $\log_a x = b$ has a unique solution.

3.3.3 How To Solve It

General 8
Given $a > 0$ and $a \neq 1$. Then $\log_a x = \log_a y$ if and only if $x = y$.

Example 45
Solve the following logarithmic equations:

1. $\log_2 x = 1$
2. $\log_3 x = 2$
3. $\log_2 x + 3 = -2\log_2 x$
4. $\log_2 3x = 1$
5. $\log_5 7x = 3$

Solution. Solve the following logarithmic equations:

1. $\log_2 x = 1$
   The equation is well-defined if and only if $x > 0$.
   We have
   $$\log_2 x = 1$$

$$\log_2 x = \log_2 2$$
$$x = 2.$$

Therefore, $x = 2$.

2. $\log_3 x = 2$
The equation is well-defined if and only if $x > 0$.
We have
$$\log_3 x = 2$$
$$\log_3 x = \log_3 3^2$$
$$x = 3^2$$
$$x = 9.$$

Therefore, $x = 9$.

3. $\log_2 x + 3 = -2\log_2 x$
The equation is well-defined if and only if $x > 0$.
We have
$$\log_2 x + 3 = -2\log_2 x$$
$$\log_2 x + 2\log_2 x = -3$$
$$3\log_2 x = -3$$
$$\log_2 x = -1$$
$$\log_2 x = \log_2 2^{-1}$$
$$x = 2^{-1}$$
$$x = \frac{1}{2}.$$

Therefore, $x = \dfrac{1}{2}$.

4. $\log_2 3x = 1$
The equation is well-defined if and only if $3x > 0$. Then $x > 0$.
We have
$$\log_2 3x = 1$$
$$\log_2 3x = \log_2 2$$

3.3. Logarithmic Equations and Inequalities:

$$3x = 2$$
$$x = \frac{2}{3}.$$

Therefore, $x = \frac{2}{3}$.

5. $\log_5 7x = 3$

The equation is well-defined if and only if $7x > 0$. Then $x > 0$.

$$\log_5 7x = 3$$
$$\log_5 7x = \log_5 3^5$$
$$7x = 243$$
$$x = \frac{243}{7}.$$

Therefore, $x = \frac{243}{7}$.

Example 46

Solve the following logarithmic equations:

1. $\log_7^2 x + \log_7 x - 2 = 0$
2. $\log_2 x + \log_x 2 = 2$
3. $(1 + \log_3 x) \log_3 x = 6$
4. $(1 + \log_4 x)(1 + 2\log_4 x) = 6$

Solution. Solve the following logarithmic equations:

1. $\log_7^2 x + \log_7 x - 2 = 0$

    The equation is well-defined if and only if $x > 0$. The given equation is a quadratic equation of variable $\log_7 x$. We see that the sum of all coefficients of the equation is zero. It turns out that $\log_7 x = 1$ or $\log_7 x = \frac{c}{a} = -\frac{2}{1} = -2$.

    - For $\log_7 x = 1$, we obtain $\log_7 x = \log_7 7$. Hence, $x = 7$.

- For $\log_7 x = -2$, we obtain $\log_7 x = \log_7 7^{-2}$.
  Then $x = \dfrac{1}{7^2} = \dfrac{1}{49}$.

$\boxed{\text{Therefore, } x \in \left\{\dfrac{1}{49}, 7\right\}.}$

2. $\log_2 x + \log_x 2 = 2$

   The equation is well-defined if and only if $\begin{cases} x > 0 \\ x \neq 1 \end{cases}$.

   We have
   $$\log_2 x + \log_x 2 = 2$$
   $$\log_2 x + \dfrac{1}{\log_2 x} = 2$$
   $$\log_2^2 x + 1 = 2\log_2 x$$
   $$\log_2^2 x - 2\log_2 x + 1 = 0$$
   $$(\log_2 x - 1)^2 = 0$$
   $$\log_2 x - 1 = 0$$
   $$\log_2 x = 1$$
   $$\log_2 x = \log_2 2.$$

$\boxed{\text{Therefore, } x = 2.}$

3. $(1 + \log_3 x)\log_3 x = 6$

   The equation is well-defined if and only if $x > 0$.
   We have
   $$\log_3 x + \log_3^2 x = 6$$
   $$\log_3^2 x + \log_3 x - 6 = 0$$
   $$(\log_3 x - 2)(\log_3 x + 3) = 0.$$

   Then $\begin{bmatrix} \log_3 x - 2 = 0 \\ \log_3 x + 3 = 0 \end{bmatrix}$. It follows that $\begin{bmatrix} \log_3 x = 2 \\ \log_3 x = -3 \end{bmatrix}$.

   - If $\log_3 x = 2$, we obtain $\log_3 x = \log_3 3^2$. Then $x = 3^2 = 9$.
   - If $\log_3 x = -3$, we obtain $\log_3 x = \log_3 3^{-3}$.
     It follows that $x = 3^{-3} = \dfrac{1}{3^3} = \dfrac{1}{27}$.

3.3. Logarithmic Equations and Inequalities:

Therefore, $x \in \left\{\dfrac{1}{27}, 9\right\}$.

4. $(1 + \log_4 x)(1 + 2\log_4 x) = 6$
The equation is well-defined if and only if $x > 0$.
We have

$$(1 + \log_4 x)(1 + 2\log_4 x) = 6$$
$$1 + 2\log_4 x + \log_4 x + 2(\log_4 x)^2 = 6$$
$$2(\log_4 x)^2 + 3\log_4 x - 5 = 0.$$

The last equation is a quadratic equation with variable $\log_4 x$. Since the sum of all coefficients of the quadratic equation is zero. We obtain $\log_4 x = 1$ or $\log_4 x = \dfrac{c}{a} = -\dfrac{5}{2}$.

- If $\log_4 x = 1$, we obtain $\log_4 x = \log_4 4$. It follows that $x = 4$.
- If $\log_4 x = -\dfrac{5}{2}$, we obtain $x = 4^{-\frac{5}{2}} = 2^{-5} = \dfrac{1}{2^5} = \dfrac{1}{32}$.

Therefore, $x \in \left\{\dfrac{1}{32}, 4\right\}$.

> **Example 47**
>
> Solve the following logarithmic equations:
>
> 1. $\log_2(x - 1) + \log_2(x + 1) = 3$
> 2. $\log_3(x + 2) - \log_3(x - 1) = \log_3 6 - 1$
> 3. $\log_2 x + \log_3 x = 1 + \log_3 2$
> 4. $\log_3 x + \log_9 x = 3$

Solution. Solve the following logarithmic equations:

1. $\log_2(x - 1) + \log_2(x + 1) = 3$
The equation is well-defined if and only if $\begin{cases} x - 1 > 0 \\ x + 1 > 0 \end{cases}$.

Then $\begin{cases} x > 1 \\ x > -1 \end{cases}$. It follows that $x > 1$.
We have

$$\log_2(x-1) + \log_2(x+1) = 3$$
$$\log_2(x-1)(x+1) = 3$$
$$\log_2(x^2 - 1) = 3$$
$$x^2 - 1 = 3$$
$$x^2 = 4.$$

We obtain $x = 2$ since $x > 1$.
$\boxed{\text{Therefore, } x = 2.}$

2. $\log_3(x+2) - \log_3(x-1) = \log_3 6 - 1$
The equation is well-defined if and only if $\begin{cases} x+2 > 0 \\ x-1 > 0 \end{cases}$.
Then $\begin{cases} x > -2 \\ x > 1 \end{cases}$.
We obtain $x > 1$.
We have

$$\log_3(x+2) - \log_3(x-1) = \log_3 6 - 1$$
$$\log_3\left(\frac{x+2}{x-1}\right) = \log_3 6 - \log_3 3$$
$$\log_3\left(\frac{x+2}{x-1}\right) = \log_3 2.$$

It follows that

$$\frac{x+2}{x-1} = 2$$
$$x+2 = 2(x-1)$$
$$x+2 = 2x - 2$$
$$2x - x = 4$$
$$x = 4.$$

$\boxed{\text{Therefore, } x = 4.}$

3. $\log_2 x + \log_3 x = 1 + \log_3 2$
The equation is well-defined if and only if $x > 0$.

68

3.3. Logarithmic Equations and Inequalities:

- If $x = 1$, the equation has no solution.
- If $x \neq 1$, we have

$$\log_2 x + \log_3 x = 1 + \log_3 2$$
$$\frac{1}{\log_x 2} + \log_3 x = 1 + \log_3 2$$
$$1 + \log_x 2 \log_3 x = (1 + \log_3 2) \log_x 2$$
$$(1 + \log_3 2) \log_x 2 = 1 + \log_3 x \log_x 2$$
$$(1 + \log_3 2) \log_x 2 = 1 + \log_3 2$$
$$\log_x 2 = 1$$
$$\log_x 2 = \log_x x.$$

Therefore, $x = 2$.

4. $\log_3 x + \log_9 x = 3$
 The equation is well-defined if and only if $x > 0$.
 We have

$$\log_3 x + \log_9 x = 3$$
$$\log_3 x + \log_{3^2} x = 3$$
$$\log_3 x + \frac{1}{2}\log_3 x = 3$$
$$\frac{3}{2}\log_3 x = 3$$
$$\log_3 x = 2$$
$$\log_3 x = \log_3 3^2.$$

Therefore, $x = 9$.

Example 48

Solve the following logarithmic equations:

1. $\log_2 \left(x^2 - 3x + 1\right) = 0$

2. $\log_4 \left(x^2 - 3x + 2\right) = \frac{1}{2} + \log_4 3$

3. $\log_3 (x - 1)(x + 2) = 1 + \log_3 2$

> 4. $\log_5\left(\dfrac{x^2-3x-1}{9}\right) = 1$

Solution. Solve the following logarithmic equations:

1. $\log_2\left(x^2 - 3x + 1\right) = 0$
   The equation is well-defined if and only if $x^2 - 3x + 1 > 0$.
   The discriminant of $x^2 - 3x + 1$ is
   $$\Delta = b^2 - 4ac$$
   $$= (-3)^2 - 4(1)(1)$$
   $$= 9 - 4$$
   $$= 5.$$

Then
$$x_1 = \frac{-b + \sqrt{\Delta}}{2a} = \frac{-(-3) + \sqrt{5}}{2(1)} = \frac{3 + \sqrt{5}}{2}$$

and
$$x_2 = \frac{-b - \sqrt{\Delta}}{2a} = \frac{-(-3) - \sqrt{5}}{2(1)} = \frac{3 - \sqrt{5}}{2}.$$

It follows that $x^2 - 3x + 1 > 0$ if and only if $x < \dfrac{3 - \sqrt{5}}{2}$ or $x > \dfrac{3 + \sqrt{5}}{2}$.
We have
$$\log_2\left(x^2 - 3x + 1\right) = 0$$
$$\log_2\left(x^2 - 3x + 1\right) = \log_2 1$$
$$x^2 - 3x + 1 = 1$$
$$x^2 - 3x = 0$$
$$x(x-3) = 0.$$

Then $\left[\begin{array}{l} x = 0 \\ x - 3 = 0 \end{array}\right.$. It follows that $\left[\begin{array}{l} x = 0 \\ x = 3 \end{array}\right.$.

Therefore, $x \in \{0, 3\}$.

3.3. Logarithmic Equations and Inequalities:

2. $\log_4 (x^2 - 3x + 2) = \dfrac{1}{2} + \log_4 3$

   The equation is well-defined if and only if $x^2 - 3x + 2 > 0$.
   If $x^2 - 3x + 2 = 0$, we obtain $x_1 = 1$ and $x_2 = \dfrac{c}{a} = \dfrac{2}{1} = 2$.
   It follows that $x^2 - 3x + 2 > 0$ if and only if $x < 1$ or $x > 2$.
   We have
   $$\log_4 (x^2 - 3x + 2) = \dfrac{1}{2} + \log_4 3$$
   $$\log_4 (x^2 - 3x + 2) = \log_4 2 + \log_4 3$$
   $$\log_4 (x^2 - 3x + 2) = \log_4 6$$
   $$x^2 - 3x + 2 = 6$$
   $$x^2 - 3x - 4 = 0.$$

   Since $a + c = 1 + (-4) = -3 = b$, we obtain $x_1 = -1$ and $x_2 = -\dfrac{c}{a} = -\dfrac{-4}{1} = 4$.

   Therefore, $x \in \{-1, 4\}$.

3. $\log_3 (x-1)(x+2) = 1 + \log_3 2$

   The equation is well-defined if and only if
   $$(x-1)(x+2) > 0.$$

   If $(x-1)(x+2) = 0$, we obtain $\begin{bmatrix} x-1 = 0 \\ x+2 = 0 \end{bmatrix}$. Then $\begin{bmatrix} x = 1 \\ x = -2 \end{bmatrix}$.
   It follows that $(x-1)(x+2) > 0$ if and only if $x < -2$ or $x > 1$.
   We have
   $$\log_3 (x-1)(x+2) = 1 + \log_3 2$$
   $$\log_3 (x-1)(x+2) = \log_3 3 + \log_3 2$$
   $$\log_3 (x-1)(x+2) = \log_3 6$$
   $$(x-1)(x+2) = 6$$
   $$x^2 + 2x - x - 2 = 6$$
   $$x^2 + x - 8 = 0.$$

   The discriminant of the last quadratic equation is
   $$\Delta = b^2 - 4ac$$

$$= 1^2 - 4(1)(-8)$$
$$= 1 + 32$$
$$= 33.$$

Then
$$x_1 = \frac{-b + \sqrt{\Delta}}{2a} = \frac{-1 + \sqrt{33}}{2(1)} = \frac{-1 + \sqrt{33}}{2}$$

and
$$x_2 = \frac{-b - \sqrt{\Delta}}{2a} = \frac{-1 - \sqrt{33}}{2(1)} = \frac{-1 - \sqrt{33}}{2}.$$

Therefore, $x \in \left\{ \dfrac{-1 - \sqrt{33}}{2}, \dfrac{-1 + \sqrt{33}}{2} \right\}$.

4. $\log_5 \left(\dfrac{x^2 - 3x - 1}{9} \right) = 1$

The equation is well-defined if and only if $\dfrac{x^2 - 3x - 1}{9} > 0$. The inequality is equivalent to $x^2 - 3x - 1 > 0$. The discriminant of $x^2 - 3x - 1$ is

$$\Delta = b^2 - 4ac$$
$$= (-3)^2 - 4(1)(-1)$$
$$= 9 + 4$$
$$= 13.$$

It follows that
$$x_1 = \frac{-b + \sqrt{\Delta}}{2a} = \frac{-(-3) + \sqrt{13}}{2(1)} = \frac{3 + \sqrt{13}}{2}$$

and
$$x_2 = \frac{-b - \sqrt{\Delta}}{2a} = \frac{-(-3) - \sqrt{13}}{2(1)} = \frac{3 - \sqrt{13}}{2}.$$

We obtain $x < \dfrac{3 - \sqrt{13}}{2}$ or $x > \dfrac{3 + \sqrt{13}}{2}$.

We have
$$\log_5 \left(\frac{x^2 - 3x - 1}{9} \right) = 1$$

3.4. Logarithmic Inequalities

$$\log_5\left(\frac{x^2 - 3x - 1}{9}\right) = \log_5 5$$

$$\frac{x^2 - 3x - 1}{9} = 5$$

$$x^2 - 3x - 1 = 45$$

$$x^2 - 3x - 46 = 0.$$

The discriminant of the last quadratic equation is

$$\Delta = (-3)^2 - 4(1)(-46)$$
$$= 9 + 184$$
$$= 193.$$

We obtain

$$x_1 = \frac{-(-3) + \sqrt{193}}{2} = \frac{3 + \sqrt{193}}{2}$$

and

$$x_2 = \frac{-(-3) - \sqrt{193}}{2} = \frac{3 - \sqrt{193}}{2}.$$

Hence, $x \in \left\{\dfrac{3 - \sqrt{193}}{2}, \dfrac{3 + \sqrt{193}}{2}\right\}$.

3.4 Logarithmic Inequalities

General 9

Suppose that a is a positive real number. We obtain the following properties:

- If $a > 1$, the inequality $\log_a f(x) < \log_a g(x)$ holds if and only if $\begin{cases} f(x) > 0 \\ f(x) < g(x) \end{cases}$.

- If $a < 1$, the inequality $\log_a f(x) < \log_a g(x)$ holds if and only if $\begin{cases} g(x) > 0 \\ f(x) > g(x) \end{cases}$.

> **Example 49**
>
> Solve the following logarithmic inequalities:
>
> 1. $\log_2(x-1) < 1$
> 2. $\log_3(2x-6) < 1$
> 3. $\log_{\frac{1}{2}}(x-1) > -1$
> 4. $\log_{\frac{1}{3}}(2x-1) < 2$

Solution. Solve the following logarithmic inequalities:

1. $\log_2(x-1) < 1$
   The given inequality is well-defined if and only if $x - 1 > 0$. It follows that $x > 1$.

   We have
   $$\log_2(x-1) < 1$$
   $$\log_2(x-1) < \log_2 2$$
   $$x - 1 < 2$$
   $$x < 3.$$

   Solution on a number line:

   > Therefore, $x \in (1, 3)$.

2. $\log_3(2x-6) < 1$
   The inequality is well-defined if and only if $2x - 6 > 0$. Then $x > 3$.

   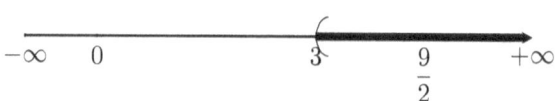

3.4. Logarithmic Inequalities

We have
$$\log_3(2x-6) < 1$$
$$\log_3(2x-6) < \log_3 3$$
$$2x - 6 < 3$$
$$2x < 9$$
$$x < \frac{9}{2}.$$

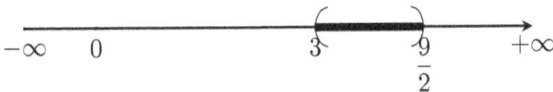

Solution on a number line:

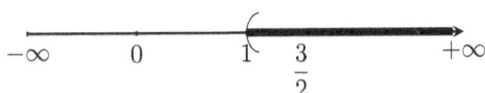

Therefore, $x \in \left(3, \dfrac{9}{2}\right)$.

3. $\log_{\frac{1}{2}}(x-1) > -1$

The inequality is well-defined if and only if $x - 1 > 0$. Then $x > 1$.

We have
$$\log_{\frac{1}{2}}(x-1) > -1$$
$$\log_{\frac{1}{2}}(x-1) > \log_{\frac{1}{2}} \frac{1}{2}$$
$$x - 1 < \frac{1}{2}$$
$$x < 1 + \frac{1}{2}$$
$$x < \frac{3}{2}.$$

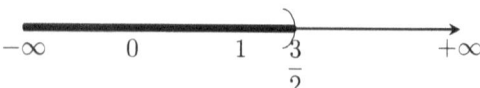

Solution on a number line:

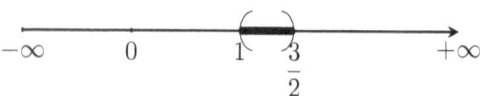

Therefore, $x \in \left(1, \dfrac{3}{2}\right)$.

4. $\log_{\frac{1}{3}}(2x-1) < 2$

   The inequality is well-defined if and only if $2x - 1 > 0$. Then $x > \dfrac{1}{2}$.

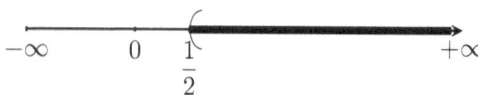

We have

$$\log_{\frac{1}{3}}(2x-1) < 2$$
$$\log_{\frac{1}{3}}(2x-1) < \log_{\frac{1}{3}}\left(\dfrac{1}{3}\right)^2$$
$$2x - 1 > \dfrac{1}{9}$$
$$2x > 1 + \dfrac{1}{9}$$
$$2x > \dfrac{10}{9}$$
$$x > \dfrac{5}{9}.$$

3.4. Logarithmic Inequalities

Solution on a number line:

$$\boxed{\text{Therefore, } x \in \left(\frac{5}{9}, +\infty\right).}$$

Example 50

Solve the following logarithmic inequalities:

1. $\log_2\left(x^2 + x - 2\right) < 2$
2. $\log_3\left(x^2 + 2x - 3\right) < 1 + \log_3 4$
3. $\log_5\left(\dfrac{x-1}{x+2}\right) \geq \log_5 \dfrac{4}{7}$
4. $\log_7\left(x^2 + 2x\right) \leq 1 + \log_7 9$

Solution. Solve the following logarithmic inequalities:

1. $\log_2\left(x^2 + x - 2\right) < 2$
   The inequality is well-defined if and only if $x^2 + x - 2 > 0$.
   If $x^2 + x - 2 = 0$, we obtain $x_1 = 1$ and $x_2 = \dfrac{c}{a} = \dfrac{-2}{1} = -2$.
   It follows that $x^2 + x - 2 > 0$ if and only if $x \in (-\infty, -2) \cup (1, +\infty)$.

   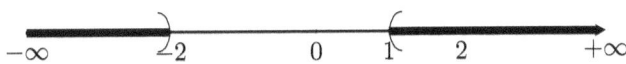

   We have

   $$\log_2\left(x^2 + x - 2\right) < 2$$
   $$\log_2\left(x^2 + x - 2\right) < \log_2 4$$
   $$x^2 + x - 2 < 4$$
   $$x^2 + x - 6 < 0.$$

If $x^2 + x - 6 = 0$, we obtain $(x+3)(x-2) = 0$. It follows that $\begin{bmatrix} x+3=0 \\ x-2=0 \end{bmatrix}$. Then $\begin{bmatrix} x=-3 \\ x=2 \end{bmatrix}$.
We obtain $x^2 + x - 6 < 0$ if and only if $-3 < x < 2$. It turns out that $x \in (-3, 2)$.

Solution on a number line:

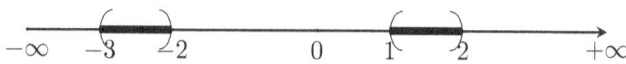

Therefore, $x \in (-3, -2) \cup (1, 2)$.

2. $\log_3 (x^2 + 2x - 3) < 1 + \log_3 4$

The inequality is well-defined if and only if $x^2 + 2x - 3 > 0$.
If $x^2 + 2x - 3 = 0$, we obtain $x_1 = 1$ and $x_2 = \dfrac{c}{a} = \dfrac{-3}{1} = -3$.
It follows that $x^2 + 2x - 3 > 0$ if and only if $x < -3$ or $x > 1$. Namely $x \in (-\infty, -3) \cup (1, +\infty)$.

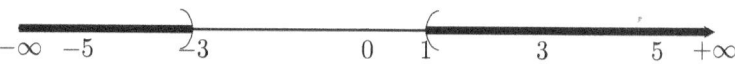

We have

$$\log_3 (x^2 + 2x - 3) < 1 + \log_3 4$$
$$\log_3 (x^2 + 2x - 3) < \log_3 3 + \log_3 4$$
$$\log_3 (x^2 + 2x - 3) < \log_3 12$$
$$x^2 + 2x - 3 < 12$$
$$x^2 + 2x - 3 - 12 < 0$$
$$x^2 + 2x - 15 < 0.$$

If $x^2 + 2x - 15 = 0$, we obtain $(x-3)(x+5) = 0$. Then $\begin{bmatrix} x-3=0 \\ x+5=0 \end{bmatrix}$. It follows that $\begin{bmatrix} x=3 \\ x=-5 \end{bmatrix}$. Then $x^2 + 2x - 15 < 0$ if and only if $-5 < x < 3$.

3.4. Logarithmic Inequalities

Solution on a number line:

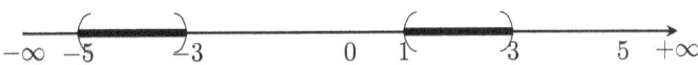

It turns out that $x \in (-5, -3) \cup (1, 3)$.

Therefore, $x \in (-5, -3) \cup (1, 3)$.

3. $\log_5 \left(\dfrac{x-1}{x+2} \right) \geq \log_5 \dfrac{4}{7}$

The inequality is well-defined if and only if $\dfrac{x-1}{x+2} > 0$.

- If $x - 1 = 0$, we obtain $x = 1$.
- If $x + 2 = 0$, we obtain $x = -2$.

Sign table of $\dfrac{x-1}{x+2}$:

x	$-\infty$		-2		1		$+\infty$
$x - 1$		$-$		$-$	0	$+$	
$x + 2$		$-$	0	$+$		$+$	
$\dfrac{x-1}{x+2}$		$+$		$-$	0	$+$	

It follows that $x \in (-\infty, -2) \cup (1, +\infty)$.

We have

$$\log_5\left(\frac{x-1}{x+2}\right) \geq \log_5 \frac{4}{7}$$

$$\frac{x-1}{x+2} - \frac{4}{7} \geq 0$$

$$\frac{7(x-1) - 4(x+2)}{7(x+2)} \geq 0$$

$$\frac{7x - 7 - 4x - 8}{7(x+2)} \geq 0$$

$$\frac{3x - 15}{7(x+2)} \geq 0.$$

- If $3x - 15 = 0$, we obtain $3x = 15$. Then $x = 5$.
- If $x + 2 = 0$, we obtain $x = -2$.

x	$-\infty$		-2		5		$+\infty$
$3x - 5$		$-$		$-$	0	$+$	
$x + 2$		$-$	0	$+$		$+$	
$\dfrac{3x-15}{x+2}$		$+$		$-$	0	$+$	

It implies that $x \in (-\infty, -2) \cup [5, +\infty)$.

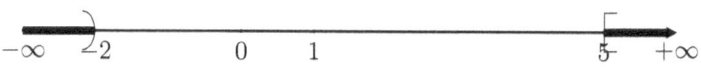

Solution on a line number:

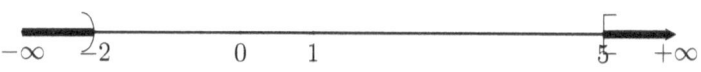

Therefore, $x \in (-\infty, -2) \cup [5, +\infty)$.

3.4. Logarithmic Inequalities

4. $\log_7 (x^2 + 2x) \leq 1 + \log_7 9$

The inequality is well-defined if and only if $x^2 + 2x > 0$. If $x^2 + 2x = 0$, we obtain $x(x+2) = 0$. Then $\left[\begin{array}{l} x = 0 \\ x + 2 = 0 \end{array} \right.$. It follows that $\left[\begin{array}{l} x = 0 \\ x = -2 \end{array} \right.$.

Consequently, $x^2 + 2x > 0$ if and only if $x < -2$ and $x > 0$. Namely, $x \in (-\infty, -2) \cup (0, +\infty)$.

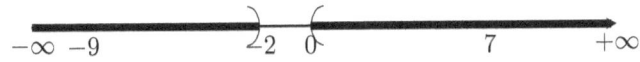

We have

$$\log_7 (x^2 + 2x) \leq 1 + \log_7 9$$
$$\log_7 (x^2 + 2x) \leq \log_7 7 + \log_7 9$$
$$\log_7 (x^2 + 2x) \leq \log_7 63$$
$$x^2 + 2x \leq 63$$
$$x^2 + 2x - 63 \leq 0.$$

If $x^2 + 2x - 63 = 0$, we obtain $(x - 7)(x + 9) = 0$. It follows that $\left[\begin{array}{l} x - 7 = 0 \\ x + 9 = 0 \end{array} \right.$. Then $\left[\begin{array}{l} x = 7 \\ x = -9 \end{array} \right.$.

We obtain $x^2 + 2x - 63 \leq 0$ if and only if $-9 \leq x \leq 7$. Then $x \in [-9, 7]$.

Solution on a number line:

Therefore, $x \in [-9, -2) \cup (0, 7]$.

Chapter 3. Logarithmic Functions

Chapter 4

Problems

Problem 1. Choose the correct answers:

1. 3^5 is equal to

   (a) 81   (b) 243   (c) 242   (d) 102

2. $(-2)^4$ is equal to

   (a) 16   (b) -16 (c) 32   (d) -32

3. $(-1)^2 + (-1)^3$ is equal to

   (a) 0   (b) -1 (c) -2 (d) 2

Problem 2. Compute the following expressions:

1. -3^0
2. $(-4)^0$
3. $\left(x^{-1}y^{-3}\right)^{-1}$
4. $2^0 + 2^{-1}$
5. $\dfrac{2^0 + 2^{-1}}{3^{-2}}$
6. $\left(\dfrac{a^2}{b^{-2}}\right)^2$
7. 4×8^n
8. $\dfrac{2^m}{16}$

Problem 3. Compute the following expressions:

1. $(0.0001)^0$
2. $(-21^{20})^0$
3. 12×1321^0
4. $\dfrac{(8^{2-2n})(16^{3-n})}{(4^{2n})^{-1}}$
5. $2\left(\dfrac{4}{m}\right)^{-2}$
6. $\left(\dfrac{6}{12m^{-3}}\right)^{-1}$
7. $\dfrac{2^{-1}+3^{-1}}{4^{-1}}$
8. $\dfrac{x^{-3}y^4}{x^4 y^{-3}}$
9. $\dfrac{m^{-1}+n^{-1}}{m^{-1}-n^{-1}}$

Problem 4. Compute the following expressions:

1. $2^{-2}+3^{-3}$
2. $5-\dfrac{1}{7^{-2}}$
3. $\dfrac{1}{3^{-2}}-\dfrac{1}{2^{-2}}$
4. $\dfrac{1}{3+2^{-1}}-\dfrac{1}{4+3^{-1}}$
5. $\left(\dfrac{1}{2+2^{-1}+2^{-2}}\right)^{-1}$

Problem 5. Compute the following expressions:

1. $(-1)^5+(-1)^2+2(-1)^4$
2. $(-1)^6-(-1)^3+(-2)^5$
3. $2(-1)^3-3(-1)^4+4(-1)^5$
4. $3(-1)^4+2(-1)^3-4(-1)^2$

Problem 6. Evaluate the following expressions:

1. $(-1)^1+(-1)^2$
2. $(-1)^1+(-1)^2+(-1)^3$
3. $(-1)^1+(-1)^2+(-1)^3+(-1)^4$
4. $(-1)^1+(-1)^2+(-1)^3+(-1)^4+(-1)^5$

Use the above result to find $(-1)^1+(-1)^2+\ldots+(-1)^n$ for all natural integers n.

Problem 7. Simplify the following sums:

1. $2^n+2^{n+1}+2^{n+2}$
2. $3^n-3^{n+1}+3^{n+2}$
3. $5^{n-1}-5^n+5^{n+1}$
4. $7^{3n}+7^{3n+1}-7^{3n+2}$

Problem 8. Simplify the following expressions:

1. $(2^x-1)(2^x+2)$

2. $(4^x - 2)(2^x + 4^x)$

3. $(1 - 2^x)(2 + 2^x)$

4. $2^x(1 - 3^x) + 3^x(2^x + 1)$

5. $(2^x - 1)(3^x + 2) - (3^x - 2)(2^x + 2)$

Problem 9. Factor the following expressions:

1. $4^x - 1$
2. $9^x - 1$
3. $8^x - 1$
4. $8^x + 1$
5. $27^x - 1$
6. $27^x + 1$
7. $8^x - 27^x$
8. $8^x + 27^x$
9. $-4^x + 9^x$
10. $-9^x + 4^x$

Problem 10. Simplify the following expressions:

1. $\sqrt{4}$
2. $\sqrt{9}$
3. $\sqrt{16}$
4. $\sqrt{25}$
5. $\sqrt{12}$
6. $\sqrt{72}$
7. $\sqrt{162}$
8. $\sqrt{468}$
9. $\sqrt{1024}$
10. $\sqrt{2048}$

Problem 11. Write the following radicals in fractional exponential forms:

1. $\sqrt{13}$
2. $\sqrt[3]{12}$
3. $\sqrt[5]{19}$
4. $\sqrt[6]{4}$
5. $\sqrt[12]{8}$
6. $\sqrt[4]{18}$
7. $\sqrt[23]{32}$
8. $\sqrt[7]{16}$

Problem 12. Solve the following exponential equations:

1. $2^x = 2$
2. $3^x = 3^{1-x}$
3. $4^x = 4^{x^2 - x}$
4. $5^{x^2 - 3x + 1} = 5$
5. $6^{x^2 - 4x + 2} = 6$
6. $7^{x^2 - 5x + 2} = 7^2$
7. $10^{\sqrt{x} - 1} = 10$
8. $11^{\frac{x-1}{x+1}} = 11$
9. $13^{x^3 - x + 1} = 13$
10. $15^{x^3 - 3x^2 + 1} = 15^{-1}$

Problem 13. Solve the following exponential equations:

1. $2^x = 16$
2. $6^x = 216$
3. $3^x = 81$
4. $0.4^x = \dfrac{25}{4}$
5. $3^x = \dfrac{9}{\sqrt[3]{9}}$
6. $0.2^x = \sqrt[3]{0.008}$
7. $\left(\dfrac{1}{2}\right)^x = \dfrac{1}{128}$
8. $\left(\dfrac{1}{4}\right)^x = \dfrac{1}{256}$
9. $3^{-x} = \dfrac{1}{243}$
10. $\left(\dfrac{2}{3}\right)^x = 1.5$
11. $\left(\dfrac{2}{7}\right)^x = \dfrac{343}{8}$
12. $\left(\dfrac{3}{4}\right)^x = \dfrac{64}{27}$
13. $25^x = 5$
14. $3^{2^x} = 6561$
15. $81^{4^x} = 9$
16. $2^x \times 5^x = 1$
17. $\left(2 - \sqrt{3}\right)^x + \left(2 + \sqrt{3}\right)^x = 2$
18. $3 \times 4^x + 2 \times 9^x - 5 \times 6^x = 0$
19. $\left(\sqrt{2 + \sqrt{3}}\right)^x + \left(\sqrt{2 - \sqrt{3}}\right)^x = 4$

Problem 14. Solve the following exponential equations:

1. $3^{2x} = 27$
2. $2^{5x} = \dfrac{1}{128}$
3. $0.5^{-2x} = 0.125$
4. $17^{3x} = 1$
5. $0.2^{7x} = 0.04$
6. $3^{x+1} = 9$
7. $8^{x-3} = 16$
8. $32^x = 16^{1-x}$
9. $9^{4x-2} = \dfrac{1}{81}$
10. $0.5^{3x-2} = 0.125$
11. $\sqrt{7^{3x+1}} = 49$
12. $4^x \times 16^{3x-1} = 8$
13. $8^{2x} \times 4^{2x-1} = 16$
14. $2^x \times 3^x = 216$
15. $5^x \times 2^x = 0.001$
16. $5^{x-1} \times 2^{x-1} = 0.001$
17. $7^{1-x} \times 4^{1-x} = \dfrac{1}{28}$
18. $\sqrt{12^x} \times \sqrt{3^x} = \dfrac{1}{6}$
19. $\sqrt{7^{x-1}} \times \sqrt{2^{x-1}} = 1$
20. $2^x + 2^{x+2} = 20$
21. $3^x + 3^{x+1} = 12$
22. $5^x + 5^{x-1} = 30$
23. $4^{x-1} + 4^x + 4^{x+1} = 84$

Problem 15. Solve the following exponential equations:

1. $3^{x^2} = 81^{x-1}$
2. $9^{(2x-1)^2} = 3^{x+3}$
3. $2^{2x+3} - 2^x = 1 - 2^{x+3}$
4. $3^{x^2+4x} = \dfrac{1}{27}$
5. $3^{5x} \times 9^{x^2} = 27$
6. $4^{3x^2+2x+1} = 16$

Problem 16. Solve the following exponential equations:

1. $9^x + 3^x - 2 = 0$
2. $16^x - 5 \times 4^x + 4 = 0$
3. $4^x - 3 \times 2^x + 2 = 0$
4. $64^x - 3 \times 2^{3x+1} + 8 = 0$
5. $9^x - 4 \times 3^{x+1} + 27 = 0$

Problem 17. Solve the following exponential equations:

1. $2^{x^2-2x+1} = 1$
2. $5^{2x+1} - 3 \times 5^{2x-1} = 550$

3. $3^{2x-1} \times 5^{3x+2} = \dfrac{9}{5} \times 5^{2x} \times 3^{3x}$

4. $4^x - 3^{x-\frac{1}{2}} = 3^{x+\frac{1}{2}} - 2^{2x-1}$

5. $3 \times 16^x + 37 \times 36^x = 26 \times 81^x$

6. $3^{2x^2+6x-9} + 4 \times 15^{x^2+3x-5} = 3 \times 5^{2x^2+6x-9}$

7. $27^x + 12^x = 2 \times 8^x$

8. $5 \times 2^{3x-3} - 3 \times 2^{5-3x} + 7 = 0$

9. $3^{2x^2-6x+3} + 6^{x^2-3x+1} = 2^{2x^2-6x+3}$

Problem 18. Solve the following exponential equations:

1. $3^x + 4^x = 5^x$
2. $3^x + 4^x + 5^x = 6^x$
3. $10^x + 11^x + 12^x = 13^x + 14^x$
4. $6^x - 8^x - 9^x + 12^x = 0$
5. $2^x + 3^x - 4^x + 6^x - 9^x = 1$

Problem 19. Solve the following exponential equations:

1. $12^x - 3^x - 4^x + 1 = 0$
2. $6^x + 2^x - 3^x - 1 = 0$
3. $20^x - 4^x + 5^x - 1 = 0$
4. $30^x - 6 \times 5^x - 6^x + 6 = 0$

Problem 20. Solve the following exponential inequalities:

1. $2^x > 4$
2. $2^x < 4$
3. $0.1^x > 100$
4. $\left(\dfrac{1}{5}\right)^x > \sqrt[3]{0.04}$
5. $0.3^x > \dfrac{100}{9}$
6. $15^x > \dfrac{1}{\sqrt[4]{15^3}}$
7. $2^x < \dfrac{1}{64}$
8. $11^x > \sqrt[5]{11}$
9. $3^x < \dfrac{1}{9\sqrt{3}}$

Problem 21. Solve the following exponential inequalities:

1. $25^{-4x} < 5\sqrt{5}$
2. $\left(\dfrac{2}{3}\right)^{-3x} < \dfrac{16}{81}$
3. $16^x > 0.125$
4. $\left(\dfrac{1}{49}\right)^{6x} > 7\sqrt[3]{49}$
5. $2^x > -5$
6. $2^x < -5$
7. $100^{x+1} < 10000$
8. $10^{x-2} \geq 0.01$
9. $27^x \times 3^{1-x} < \dfrac{1}{3}$

Problem 22. Solve the following exponential inequalities:

1. $0.1^x > 10$
2. $12^x < \sqrt[3]{144}$
3. $0.4^x < \dfrac{125}{8}$
4. $2^x > \sqrt{2^3}$
5. $10^x > \dfrac{1}{\sqrt{10}}$
6. $0.2^x < 25$
7. $5^x < \sqrt[3]{25}$
8. $0.7^x > \dfrac{100}{49}$
9. $\left(\dfrac{1}{3}\right)^x > \sqrt[3]{\dfrac{1}{9}}$

Problem 23. Solve the following exponential inequalities:

1. $25^x > 125^{3x-2}$
2. $4^{-x+\frac{1}{2}} - 7 \times 2^{-x} - 4 < 0$
3. $5^{2x-1} \times 7^{3x+2} \leq 7^{2x-1} \times 5^{3x+2}$
4. $2^{x+2} - 2^{x+3} - 2^{x+4} > 5^{x+1} - 5^{x+2}$
5. $0.1^{4x^2-2x-2} \leq 0.1^{2x-3}$
6. $2^{x^2} \times 5^{x^2} < 10^{-3}\left(10^{3-x}\right)^2$
7. $2^{9x-x^3} > 1$
8. $2^{9x-x^3} < 1$
9. $2^{2x^2-6x+3} + 6^{x^2-3x+1} \geq 3^{2x^2-6x+3}$

Problem 24. By definition, write the following expressions in logarithmic forms:

1. $4^3 = 64$
2. $3^5 = 243$
3. $7^3 = 343$
4. $10^{-2} = \dfrac{1}{100}$
5. $4^{-3} = \dfrac{1}{64}$
6. $x^y = k$

Problem 25. Compute the following expressions:

1. $2^{\log_2 5}$
2. $2^{2\log_2 3}$
3. $\log_3 27$
4. $\log_{\sqrt{2}} 32$
5. $\log_{\sqrt{2}} \sqrt{8}$
6. $\log 0.01$
7. $\log_{2022} 1$
8. $\log_{\sqrt[4]{3}} \sqrt[4]{27}$
9. $\log 2 + \log 5$
10. $\log_2 10 - \log_2 5$

Problem 26. Given three positive real numbers $a, b, x \neq 1$ such that $\log_a x = y$ and $\log_b x = z$. Find $\log_{ab} x$ in terms of y and z.

Problem 27. Prove that $\log_2 \log_3 \sqrt{3\sqrt{3\sqrt{3\sqrt{3}}}} = \log_2 15 - 4$.

Problem 28. Compute $3^{2\log_3 2} + 4^{3\log_2 5} + 5^{\frac{1}{\log_2 5}}$.

Problem 29. Given a and b are two positive numbers greater than 1 such that $\log_a b = 2$. Compute

1. $\log_{ab}\left(\dfrac{\sqrt[4]{a}}{\sqrt[3]{b}}\right)$
2. $\log_{\sqrt[3]{a}\sqrt[4]{b}}\left(\dfrac{a}{b}\right)$

Problem 30. Let a and b be two positive real numbers such that $a^2 + b^2 = 3ab$. Prove that $\log_5(a+b) = \dfrac{1}{2}(1 + \log_5 a + \log_5 b)$.

Problem 31. For all $x > 1$, prove that

$$\dfrac{1}{\log_2 x} + \dfrac{1}{\log_3 x} + \ldots + \dfrac{1}{\log_{2022} x} = \log_x 2022!.$$

Problem 32. Given a, b and $c > 1$ such that $\log_a b + \log_b c + \log_c a = 1$. Compute

$$\dfrac{\log_a c}{\log_{ab} c} + \dfrac{\log_b a}{\log_{bc} a} + \dfrac{\log_c b}{\log_{ca} b}.$$

Problem 33. Given $\log_2 3 = x$ and $\log_3 4 = y$.
Compute $E = \dfrac{1+xy}{1-xy}$.

Problem 34. Given a, b and $c > 1$. Prove that $\log_a b \times \log_b c = \log_a c$. Using induction, show that $\log_{x_1} x_2 \times \log_{x_2} x_3 \times ... \times \log_{x_{n-1}} x_n = \log_{x_1} x_n$ for all $n \geq 2$.

Problem 35. Given four real numbers a, b, c and $x > 1$. Prove that
$$\log_a x \log_b x + \log_b x \log_c x + \log_c x \log_a x = \frac{\log_a x \log_b x \log_c x}{\log_{abc} x}.$$

Problem 36. Suppose that x, y and z are three distinct positive real numbers greater than 1. If $\log_y x \log_z x + \log_x y \log_z y + \log_y z \log_x z = 3$, find xyz.

Problem 37. Given a and b are two positive real numbers such that $\log_{16} a + \log_8 b = \dfrac{1}{4}$ and $\log_{16} b + \log_8 a = \dfrac{1}{5}$. Compute ab.

Problem 38. Given x, y and z are three positive real number greater than 1 such that $a^x = b^y = c^z = k$. Prove that
$$x\log_x a + y\log_y b + z\log_z c = \frac{\log_k x \log_k y + \log_k y \log_k z + \log_k z \log_k x}{\log_k x \log_k y \log_k z}.$$

Problem 39. Given three positive real numbers a, b and $c > 1$. Prove that $a^{\log_b c} = c^{\log_b a}$.

Problem 40. Solve the following logarithmic equations:

1. $\log_2 (3x) = 1$
2. $\log_2 x + \log_2 4 = \log_2 5$
3. $3^{2\log_3 x} - 3x + 2 = 0$
4. $3\log_2 x - 1 = \log_2 5$
5. $2 + \log_3 (x-1) = \log_2 3$

Problem 41. Solve the following logarithmic equations:

1. $\log_x 8 + \log_8 x = 2$
2. $2\log_x 2 + \log_2 x = 3$
3. $2\log_x 3 + \log_9 x = 2$
4. $\log_2 (x^2 + 4x - 3) = 1$
5. $\log_3 (x^2 + 4x + 1) = \log_3 2 + 1$

Problem 42. Solve the following logarithmic inequalities:

1. $\log_2(2x) > 0$

2. $\log_3(x-1) > 1$

3. $\log_5(2x-3) < 1$

4. $\log_2(2x-1) \geq 1$

5. $\log_3(-x+1) \leq 2$

Chapter 5

Solutions

Problem 1. Choose the correct answers:

1. 3^5 is equal to

   (a) 81   (b) 243   (c) 242   (d) 102

2. $(-2)^4$ is equal to

   (a) 16   (b) -16 (c) 32   (d) -32

3. $(-1)^2 + (-1)^3$ is equal to

   (a) 0   (b) -1 (c) -2 (d) 2

Solution. Choose the correct answers:

1. 3^5 is equal to

   (a) 81   (b) $\boxed{243}$ (c) 242   (d) 102

   (b) is the correct answer since
   $$3^5 = 3 \times 3 \times 3 \times 3 \times 3 = 243.$$

2. $(-2)^4$ is equal to

(a) $\boxed{16}$ (b) -16 (c) 32 (d) -32

(a) is the correct answer since
$$(-2)^4 = (-2)(-2)(-2)(-2) = 16.$$

3. $(-1)^2 + (-1)^3$ is equal to

(a) $\boxed{0}$ (b) -1 (c) -2 (d) 2

(a) is correct since $(-1)^2 + (-1)^3 = 1 - 1 = 0$.

Problem 2. Compute the following expressions:

1. -3^0
2. $(-4)^0$
3. $\left(x^{-1}y^{-3}\right)^{-1}$
4. $2^0 + 2^{-1}$
5. $\dfrac{2^0 + 2^{-1}}{3^{-2}}$
6. $\left(\dfrac{a^2}{b^{-2}}\right)^2$
7. 4×8^n
8. $\dfrac{2^m}{16}$

Solution. 1. $-3^0 = -1$

2. $(-4)^0 = 1$

3. $\left(x^{-1}y^{-3}\right)^{-1} = \left(x^{-1}\right)^{-1}\left(y^{-3}\right)^{-1} = x^{(-1)(-1)}y^{(-3)(-1)} = xy^3$

4. $2^0 + 2^{-1} = 1 + \dfrac{1}{2} = \dfrac{2+1}{2} = \dfrac{3}{2}$

5. $\dfrac{2^0 + 2^{-1}}{3^{-2}} = \dfrac{1 + \frac{1}{2}}{\frac{1}{3^2}} = \dfrac{\frac{2+1}{2}}{\frac{1}{9}} = \dfrac{\frac{3}{2}}{\frac{1}{9}} = \dfrac{3}{2} \times \dfrac{9}{1} = \dfrac{27}{2}$

6. $\left(\dfrac{a^2}{b^{-2}}\right)^2 = \left(a^2 b^2\right)^2 = \left(a^2\right)^2\left(b^2\right)^2 = a^{2\times 2}b^{2\times 2} = a^4 b^4$

7. $4 \times 8^n = 2^2 \times \left(2^3\right)^n = 2^2 \times 2^{3n} = 2^{2+3n} = 2^{3n+2}$

8. $\dfrac{2^m}{16} = \dfrac{2^m}{2^4} = 2^{m-4}$

Problem 3. Compute the following expressions:

1. $(0.0001)^0$

2. $\left(-21^{20}\right)^0$

3. 12×1321^0

4. $\dfrac{\left(8^{2-2n}\right)\left(16^{3-n}\right)}{\left(4^{2n}\right)^{-1}}$

5. $2\left(\dfrac{4}{m}\right)^{-2}$

6. $\left(\dfrac{6}{12m^{-3}}\right)^{-1}$

7. $\dfrac{2^{-1}+3^{-1}}{4^{-1}}$

8. $\dfrac{x^{-3}y^4}{x^4 y^{-3}}$

9. $\dfrac{m^{-1}+n^{-1}}{m^{-1}-n^{-1}}$

Solution. 1. $(0.0001)^0 = 1$

2. $\left(-21^{20}\right)^0 = 1$

3. $12 \times 1321^0 = 12 \times 1 = 12$

4. $\dfrac{\left(8^{2-2n}\right)\left(16^{3-n}\right)}{\left(4^{2n}\right)^{-1}}$

We have

$$\dfrac{\left(8^{2-2n}\right)\left(16^{3-n}\right)}{\left(4^{2n}\right)^{-1}} = \dfrac{\left[(2^3)^{2-2n}\right]\left[(2^4)^{3-n}\right]}{\left[(2^2)^{2n}\right]^{-1}}$$

$$= \dfrac{2^{3(2-2n)} \times 2^{4(3-n)}}{(2^{2 \times 2n})^{-1}}$$

$$= \dfrac{2^{6-6n} \times 2^{12-4n}}{(2^{4n})^{-1}}$$

$$= 2^{6-6n} \times 2^{12-4n} \times 2^{4n}$$

$$= 2^{6-6n+12-4n+4n}$$

$$= 2^{18-6n}.$$

5. $2\left(\dfrac{4}{m}\right)^{-2} = 2 \times \dfrac{1}{\left(\dfrac{4}{m}\right)^2} = 2 \times \dfrac{1}{\dfrac{4^2}{m^2}} = 2 \times \dfrac{m^2}{4^2} = 2 \times \dfrac{m^2}{16} = \dfrac{m^2}{8}$

6. $\left(\dfrac{6}{12m^{-3}}\right)^{-1} = \left(\dfrac{1}{2m^{-3}}\right)^{-1}$
$= \left[(2m^{-3})^{-1}\right]^{-1}$
$= 2m^{-3}$
$= 2\left(\dfrac{1}{m^3}\right)$
$= \dfrac{2}{m^3}$

7. $\dfrac{2^{-1}+3^{-1}}{4^{-1}} = \dfrac{\frac{1}{2}+\frac{1}{3}}{\frac{1}{4}} = \dfrac{\frac{3+2}{6}}{\frac{1}{4}} = \dfrac{\frac{5}{6}}{\frac{1}{4}} = \dfrac{5}{6} \times 4 = \dfrac{5}{3} \times 2 = \dfrac{10}{3}$

8. $\dfrac{x^{-3}y^4}{x^4 y^{-3}} = x^{-3-4} y^{4-(-3)} = x^{-7} y^7 = \left(\dfrac{1}{x^7}\right) y^7 = \dfrac{y^7}{x^7} = \left(\dfrac{y}{x}\right)^7$

9. $\dfrac{m^{-1}+n^{-1}}{m^{-1}-n^{-1}} = \dfrac{\frac{1}{m}+\frac{1}{n}}{\frac{1}{m}-\frac{1}{n}}$
$= \dfrac{\frac{n+m}{mn}}{\frac{n-m}{mn}}$
$= \dfrac{n+m}{mn} \times \dfrac{mn}{n-m}$
$= \dfrac{n+m}{n-m}$

Problem 4. Compute the following expressions:

1. $2^{-2} + 3^{-3}$
2. $5 - \dfrac{1}{7^{-2}}$
3. $\dfrac{1}{3^{-2}} - \dfrac{1}{2^{-2}}$
4. $\dfrac{1}{3+2^{-1}} - \dfrac{1}{4+3^{-1}}$
5. $\left(\dfrac{1}{2+2^{-1}+2^{-2}}\right)^{-1}$

Solution. 1. $2^{-2} + 3^{-3} = \dfrac{1}{2^2} + \dfrac{1}{3^3} = \dfrac{1}{4} + \dfrac{1}{27} = \dfrac{27+4}{108} = \dfrac{31}{108}$

2. $5 - \dfrac{1}{7^{-2}} = 5 - 7^2 = 5 - 49 = -44$

3. $\dfrac{1}{3^{-2}} - \dfrac{1}{2^{-2}} = 3^2 - 2^2 = 9 - 4 = 5$

4. $\dfrac{1}{3 + 2^{-1}} - \dfrac{1}{4 + 3^{-1}} = \dfrac{1}{3 + \dfrac{1}{2}} - \dfrac{1}{4 + \dfrac{1}{3}}$

$= \dfrac{1}{\dfrac{6+1}{2}} - \dfrac{1}{\dfrac{12+1}{2}}$

$= \dfrac{2}{7} - \dfrac{2}{13}$

$= \dfrac{26 - 14}{91}$

$= \dfrac{12}{91}$

5. $\left(\dfrac{1}{2 + 2^{-1} + 2^{-2}}\right)^{-1} = \left(\dfrac{1}{2 + \dfrac{1}{2} + \dfrac{1}{2^2}}\right)^{-1}$

$= \left(\dfrac{1}{2 + \dfrac{1}{2} + \dfrac{1}{4}}\right)^{-1}$

$= \left(\dfrac{1}{\dfrac{8 + 2 + 1}{4}}\right)^{-1}$

$= \left(\dfrac{4}{11}\right)^{-1}$

$= \dfrac{11}{4}$

Problem 5. Compute the following expressions:

1. $(-1)^5 + (-1)^2 + 2(-1)^4$ 3. $2(-1)^3 - 3(-1)^4 + 4(-1)^5$
2. $(-1)^6 - (-1)^3 + (-2)^5$ 4. $3(-1)^4 + 2(-1)^3 - 4(-1)^2$

Solution. Compute the following expressions:
1. $(-1)^5 + (-1)^2 + 2(-1)^4$
   We have already known that $(-1)^n = 1$ if n is an even number

and $(-1)^n = -1$ if n is an odd number. It follows that
$$(-1)^5 + (-1)^2 + 2(-1)^4 = -1 + 1 + 2\,(1)$$
$$= 2.$$

Therefore, $(-1)^5 + (-1)^2 + 2(-1)^4 = 2.$

2. $(-1)^6 - (-1)^3 + (-2)^5$
We have
$$(-1)^6 - (-1)^3 + (-2)^5 = 1 - (-1) - 2^5$$
$$= 2 - 32$$
$$= -30.$$

Therefore, $(-1)^6 - (-1)^3 + (-2)^5 = -30.$

3. $2(-1)^3 - 3(-1)^4 + 4(-1)^5$
We have
$$2(-1)^3 - 3(-1)^4 + 4(-1)^5 = 2\,(-1) - 3\,(1) + 4\,(-1)$$
$$= -2 - 3 - 4$$
$$= -9.$$

Therefore, $2(-1)^3 - 3(-1)^4 + 4(-1)^5 = -9.$

4. $3(-1)^4 + 2(-1)^3 - 4(-1)^2$
We have
$$3(-1)^4 + 2(-1)^3 - 4(-1)^2 = 3\,(1) + 2\,(-1) - 4\,(1)$$
$$= 3 - 2 - 4$$
$$= -3.$$

Therefore, $3(-1)^4 + 2(-1)^3 - 4(-1)^2 = -3.$

Problem 6. Evaluate the following expressions:

1. $(-1)^1 + (-1)^2$
2. $(-1)^1 + (-1)^2 + (-1)^3$

3. $(-1)^1 + (-1)^2 + (-1)^3 + (-1)^4$

4. $(-1)^1 + (-1)^2 + (-1)^3 + (-1)^4 + (-1)^5$

Use the above result to find $(-1)^1 + (-1)^2 + ... + (-1)^n$ for all natural integers n.

Solution.

1. $(-1)^1 + (-1)^2$
   We have $(-1)^1 + (-1)^2 = -1 + 1 = 0$.
   $\boxed{\text{Therefore, } (-1)^1 + (-1)^2 = 0.}$

2. $(-1)^1 + (-1)^2 + (-1)^3$
   We have $(-1)^1 + (-1)^2 + (-1)^3 = -1 + 1 - 1 = -1$.
   $\boxed{\text{Therefore, } (-1)^1 + (-1)^2 + (-1)^3 = -1.}$

3. $(-1)^1 + (-1)^2 + (-1)^3 + (-1)^4$
   We have
   $$(-1)^1 + (-1)^2 + (-1)^3 + (-1)^4$$
   $$= -1 + 1 - 1 + 1$$
   $$= 0.$$
   $\boxed{\text{Therefore, } (-1)^1 + (-1)^2 + (-1)^3 + (-1)^4 = 0.}$

4. $(-1)^1 + (-1)^2 + (-1)^3 + (-1)^4 + (-1)^5$
   We have
   $$(-1)^1 + (-1)^2 + (-1)^3 + (-1)^4 + (-1)^5$$
   $$= -1 + 1 - 1 + 1 - 1$$
   $$= -1.$$
   $\boxed{\text{Therefore, } (-1)^1 + (-1)^2 + (-1)^3 + (-1)^4 + (-1)^5 = -1.}$

From the above patterns, we have

- $(-1)^1 + (-1)^2 + ... + (-1)^n = 0$ if n is an even number.
- $(-1)^1 + (-1)^2 + ... + (-1)^n = -1$ if n is an odd number.

Problem 7. Simplify the following sums:

Chapter 5. Solutions

1. $2^n + 2^{n+1} + 2^{n+2}$
2. $3^n - 3^{n+1} + 3^{n+2}$
3. $5^{n-1} - 5^n + 5^{n+1}$
4. $7^{3n} + 7^{3n+1} - 7^{3n+2}$

Solution. Simplify the following sums:

1. $2^n + 2^{n+1} + 2^{n+2}$
   We have
   $$2^n + 2^{n+1} + 2^{n+2} = 2^n \left(1 + 2 + 2^2\right)$$
   $$= 7 \times 2^n.$$

   $\boxed{\text{Therefore, } 2^n + 2^{n+1} + 2^{n+2} = 7 \times 2^n.}$

2. $3^n - 3^{n+1} + 3^{n+2}$
   We have
   $$3^n - 3^{n+1} + 3^{n+2} = 3^n \left(1 - 3 + 3^2\right)$$
   $$= 7 \times 3^n.$$

   $\boxed{\text{Therefore, } 3^n - 3^{n+1} + 3^{n+2} = 7 \times 3^n.}$

3. $5^{n-1} - 5^n + 5^{n+1}$
   We have
   $$5^{n-1} - 5^n + 5^{n+1} = 5^{n-1} \left(1 - 5 + 5^2\right)$$
   $$= 5^{n-1} \left(1 - 5 + 25\right)$$
   $$= 21 \times 5^{n-1}.$$

   $\boxed{\text{Therefore, } 5^{n-1} - 5^n + 5^{n+1} = 21 \times 5^{n-1}.}$

4. $7^{3n} + 7^{3n+1} - 7^{3n+2}$
   We have
   $$7^{3n} + 7^{3n+1} - 7^{3n+2} = 7^{3n} \left(1 + 7 - 7^2\right)$$
   $$= 7^{3n} \left(1 + 7 - 49\right)$$
   $$= -41 \times 7^{3n}.$$

   $\boxed{\text{Therefore, } 7^{3n} + 7^{3n+1} - 7^{3n+2} = -41 \times 7^{3n}.}$

Problem 8. Simplify the following expressions:

1. $(2^x - 1)(2^x + 2)$
2. $(4^x - 2)(2^x + 4^x)$
3. $(1 - 2^x)(2 + 2^x)$
4. $2^x(1 - 3^x) + 3^x(2^x + 1)$
5. $(2^x - 1)(3^x + 2) - (3^x - 2)(2^x + 2)$

Solution. Simplify the following expressions:

1. $(2^x - 1)(2^x + 2)$
   We have
   $$(2^x - 1)(2^x + 2) = 2^{x+x} + 2 \times 2^x - 2^x - 2$$
   $$= 2^{2x} + 2^x - 2.$$

   Therefore, $(2^x - 1)(2^x + 2) = 2^{2x} + 2^x - 2.$

2. $(4^x - 2)(2^x + 4^x)$
   We have
   $$(4^x - 2)(2^x + 4^x) = (2^{2x} - 2)(2^x + 2^{2x})$$
   $$= 2^{2x+x} + 2^{2x+2x} - 2^{x+1} - 2^{2x+1}$$
   $$= 2^{3x} + 2^{4x} - 2^{x+1} - 2^{2x+1}.$$

   Therefore, $(4^x - 2)(2^x + 4^x) = 2^{3x} + 2^{4x} - 2^{x+1} - 2^{2x+1}.$

3. $(1 - 2^x)(2 + 2^x)$
   We have
   $$(1 - 2^x)(2 + 2^x) = 2 + 2^x - 2 \times 2^x - 2^{2x}$$
   $$= 2 - 2^x - 2^{2x}$$
   $$= -2^{2x} - 2^x + 2.$$

   Therefore, $(1 - 2^x)(2 + 2^x) = -2^{2x} - 2^x + 2.$

4. $2^x(1 - 3^x) + 3^x(2^x + 1)$
   We have
   $$2^x(1 - 3^x) + 3^x(2^x + 1) = 2^x - 6^x + 6^x + 3^x$$
   $$= 2^x + 3^x.$$

   Therefore, $2^x(1 - 3^x) + 3^x(2^x + 1) = 2^x + 3^x.$

5. $(2^x - 1)(3^x + 2) - (3^x - 2)(2^x + 2)$
We have

$$(2^x - 1)(3^x + 2) - (3^x - 2)(2^x + 2)$$
$$= 6^x + 2^{x+1} - 3^x - 2 - (6^x + 2 \times 3^x - 2^{x+1} - 4)$$
$$= 6^x + 2^{x+1} - 3^x - 2 - 6^x - 2 \times 3^x + 2^{x+1} + 4$$
$$= 2 \times 2^{x+1} - 3 \times 3^x + 2$$
$$= 2^{x+2} - 3^{x+1} + 2.$$

Therefore, $(2^x - 1)(3^x + 2) - (3^x - 2)(2^x + 2) = 2^{x+2} - 3^{x+1} + 2.$

Problem 9. Factor the following expressions:

1. $4^x - 1$
2. $9^x - 1$
3. $8^x - 1$
4. $8^x + 1$
5. $27^x - 1$
6. $27^x + 1$
7. $8^x - 27^x$
8. $8^x + 27^x$
9. $-4^x + 9^x$
10. $-9^x + 4^x$

Solution. Factor the following expressions:

1. $4^x - 1$
We have

$$4^x - 1 = (2^2)^x - 1$$
$$= (2^x)^2 - 1^2$$
$$= (2^x - 1)(2^x + 1).$$

Therefore, $4^x - 1 = (2^x - 1)(2^x + 1).$

2. $9^x - 1$
We have

$$9^x - 1 = (3^2)^x - 1^2$$
$$= (3^x)^2 - 1^2$$
$$= (3^x - 1)(3^x + 1).$$

Therefore, $9^x - 1 = (3^x - 1)(3^x + 1).$

3. $8^x - 1$

    We have
    $$\begin{aligned} 8^x - 1 &= \left(2^3\right)^x - 1^3 \\ &= (2^x)^3 - 1^3 \\ &= (2^x - 1)\left[(2^x)^2 + 2^x + 1\right] \\ &= (2^x - 1)\left(2^{2x} + 2^x + 1\right). \end{aligned}$$

    Therefore, $8^x - 1 = (2^x - 1)\left(2^{2x} + 2^x + 1\right).$

4. $8^x + 1$

    We have
    $$\begin{aligned} 8^x + 1 &= \left(2^3\right)^x + 1 \\ &= (2^x)^3 + 1 \\ &= (2^x + 1)\left[(2^x)^2 - 2^x + 1\right] \\ &= (2^x + 1)\left(2^{2x} - 2^x + 1\right). \end{aligned}$$

    Therefore, $8^x + 1 = (2^x + 1)\left(2^{2x} - 2^x + 1\right).$

5. $27^x - 1$

    We have
    $$\begin{aligned} 27^x - 1 &= \left(3^3\right)^x - 1 \\ &= (3^x)^3 - 1^3 \\ &= (3^x - 1)\left[(3^x)^2 + 3^x + 1\right] \\ &= (3^x - 1)\left(3^{2x} + 3^x + 1\right). \end{aligned}$$

    Therefore, $27^x - 1 = (3^x - 1)\left(3^{2x} + 3^x + 1\right).$

6. $27^x + 1$

    We have
    $$\begin{aligned} 27^x + 1 &= \left(3^3\right)^x + 1 \\ &= (3^x)^3 + 1^3 \\ &= (3^x + 1)\left[(3^x)^2 - 3^x + 1\right] \end{aligned}$$

$$= (3^x + 1)\left(3^{2x} - 3^x + 1\right).$$

Therefore, $27^x + 1 = (3^x + 1)\left(3^{2x} - 3^x + 1\right).$

7. $8^x - 27^x$
   We have
   $$\begin{aligned} 8^x - 27^x &= \left(2^3\right)^x - \left(3^3\right)^x \\ &= (2^x)^3 - (3^x)^3 \\ &= (2^x - 3^x)\left[(2^x)^2 + (2^x)(3^x) + (3^x)^2\right] \\ &= (2^x - 3^x)\left[2^{2x} + (2 \times 3)^x + 3^{2x}\right] \\ &= (2^x - 3^x)\left(2^{2x} + 6^x + 3^{2x}\right). \end{aligned}$$

Therefore, $8^x - 27^x = (2^x - 3^x)\left(2^{2x} + 6^x + 3^{2x}\right).$

8. $8^x + 27^x$
   We have
   $$\begin{aligned} 8^x + 27^x &= \left(2^3\right)^x + \left(3^3\right)^x \\ &= (2^x)^3 + (3^x)^3 \\ &= (2^x + 3^x)\left[(2^x)^2 - (2^x)(3^x) + (3^x)^2\right] \\ &= (2^x + 3^x)\left(2^{2x} - 6^x + 3^{2x}\right). \end{aligned}$$

Therefore, $8^x + 27^x = (2^x + 3^x)\left(2^{2x} - 6^x + 3^{2x}\right).$

9. $-4^x + 9^x$
   We have
   $$\begin{aligned} -4^x + 9^x &= 9^x - 4^x \\ &= \left(3^2\right)^x - \left(2^2\right)^x \\ &= (3^x)^2 - (2^x)^2 \\ &= (3^x - 2^x)(3^x + 2^x). \end{aligned}$$

Therefore, $-4^x + 9^x = (3^x - 2^x)(3^x + 2^x).$

10. $-9^x + 4^x$
    We have
    $$-9^x + 4^x = 4^x - 9^x$$

104

$$= (2^2)^x - (3^2)^x$$
$$= (2^x)^2 - (3^x)^2$$
$$= (2^x - 3^x)(2^x + 3^x).$$

Therefore, $-9^x + 4^x = (2^x - 3^x)(2^x + 3^x).$

Problem 10. Simplify the following expressions:

1. $\sqrt{4}$
2. $\sqrt{9}$
3. $\sqrt{16}$
4. $\sqrt{25}$

5. $\sqrt{12}$
6. $\sqrt{72}$
7. $\sqrt{162}$
8. $\sqrt{468}$

9. $\sqrt{1024}$
10. $\sqrt{2048}$

Solution. Simplify the following expressions:

1. $\sqrt{4} = \sqrt{2^2} = 2$
2. $\sqrt{9} = \sqrt{3^2} = 3$
3. $\sqrt{16} = \sqrt{4^2} = 4$
4. $\sqrt{25} = \sqrt{5^2} = 5$
5. $\sqrt{12} = \sqrt{2^2 \times 3} = 2\sqrt{3}$
6. $\sqrt{72} = \sqrt{6^2 \times 2} = 6\sqrt{2}$
7. $\sqrt{162} = \sqrt{2 \times 9^2} = 9\sqrt{2}$
8. $\sqrt{468} = \sqrt{6^2 \times 13} = 6\sqrt{13}$
9. $\sqrt{1024} = \sqrt{32^2} = 32$
10. $\sqrt{2048} = \sqrt{2 \times 32^2} = 32\sqrt{2}$

Problem 11. Write the following radicals in fractional exponential forms:

1. $\sqrt{13}$
2. $\sqrt[3]{12}$
3. $\sqrt[5]{19}$

4. $\sqrt[6]{4}$
5. $\sqrt[12]{8}$
6. $\sqrt[4]{18}$

7. $\sqrt[23]{32}$
8. $\sqrt[7]{16}$

Chapter 5. Solutions

Solution. Write the following radicals in fractional exponential forms:

1. $\sqrt{13} = 13^{\frac{1}{2}}$
2. $\sqrt[3]{12} = 12^{\frac{1}{3}}$
3. $\sqrt[5]{19} = 19^{\frac{1}{5}}$
4. $\sqrt[6]{4} = 4^{\frac{1}{6}}$
5. $\sqrt[12]{8} = 8^{\frac{1}{12}}$
6. $\sqrt[4]{18} = 18^{\frac{1}{4}}$
7. $\sqrt[23]{32} = 32^{\frac{1}{23}}$
8. $\sqrt[7]{16} = 16^{\frac{1}{7}}$

Problem 12. Solve the following exponential equations:

1. $2^x = 2$
2. $3^x = 3^{1-x}$
3. $4^x = 4^{x^2-x}$
4. $5^{x^2-3x+1} = 5$
5. $6^{x^2-4x+2} = 6$
6. $7^{x^2-5x+2} = 7^2$
7. $10^{\sqrt{x}-1} = 10$
8. $11^{\frac{x-1}{x+1}} = 11$
9. $13^{x^3-x+1} = 13$
10. $15^{x^3-3x^2+1} = 15^{-1}$

Solution. Solve the following exponential equations:

1. $2^x = 2$
   Since $2^x = 2$, we obtain $x = 1$.
   $\boxed{\text{Therefore, } x = 1.}$

2. $3^x = 3^{1-x}$
   We have $3^x = 3^{1-x}$. Then $x = 1 - x$.
   It follows that $2x = 1$. Hence, $x = \dfrac{1}{2}$.
   $\boxed{\text{Therefore, } x = \dfrac{1}{2}.}$

3. $4^x = 4^{x^2-x}$
   We have $4^x = 4^{x^2-x}$. Then $x = x^2 - x$.
   It follows that $x^2 - 2x = 0$. We obtain $x(x-2) = 0$.
   Then $\begin{bmatrix} x = 0 \\ x - 2 = 0 \end{bmatrix}$. Hence, $\begin{bmatrix} x = 0 \\ x = 2 \end{bmatrix}$.
   $\boxed{\text{Therefore, } x \in \{0, 2\}.}$

4. $5^{x^2-3x+1} = 5$
   We have $5^{x^2-3x+1} = 5$. Then $x^2 - 3x + 1 = 1$.
   It follows that $x^2 - 3x = 0$. We obtain $x(x-3) = 0$.
   It implies that $\begin{bmatrix} x = 0 \\ x - 3 = 0 \end{bmatrix}$. Then $\begin{bmatrix} x = 0 \\ x = 3 \end{bmatrix}$.
   $\boxed{\text{Therefore, } x \in \{0, 3\}.}$

5. $6^{x^2-4x+2} = 6$
   We have $6^{x^2-4x+2} = 6$. Then $x^2 - 4x + 2 = 1$. It follows that $x^2 - 4x + 1 = 0$. The discriminant of the last quadratic equation is
   $$\Delta' = (b')^2 - ac$$
   $$= (-2)^2 - (1)(1)$$
   $$= 4 - 1$$
   $$= 3.$$

   Then
   $$x_1 = \frac{-b' + \sqrt{\Delta'}}{a} = \frac{2 + \sqrt{3}}{1} = 2 + \sqrt{3}$$
   and
   $$x_2 = \frac{-b' - \sqrt{\Delta'}}{a} = \frac{2 - \sqrt{3}}{1} = 2 - \sqrt{3}.$$
   $\boxed{\text{Therefore, } x \in \{2 - \sqrt{3}, 2 + \sqrt{3}\}.}$

6. $7^{x^2-5x+2} = 7^2$
   We have $7^{x^2-5x+2} = 7^2$. Then $x^2 - 5x + 2 = 2$.
   It follows that $x(x-5) = 0$. It implies that $\begin{bmatrix} x = 0 \\ x - 5 = 0 \end{bmatrix}$.
   Hence, $\begin{bmatrix} x = 0 \\ x = 5 \end{bmatrix}$.
   $\boxed{\text{Therefore, } x \in \{0, 5\}.}$

7. $10^{\sqrt{x}-1} = 10$
   The equation is well-defined if and only if $x \geq 0$. We have $10^{\sqrt{x}-1} = 10$. Then $\sqrt{x} - 1 = 1$. We obtain $\sqrt{x} = 2$. Squaring both sides of the last equation, we obtain $x = 4$.
   $\boxed{\text{Therefore, } x = 4.}$

8. $11^{\frac{x-1}{x+1}} = 11$

We have $11^{\frac{x-1}{x+1}} = 11$. Then $\dfrac{x-1}{x+1} = 1$. It follows that $x - 1 = x + 1$. Hence, $0 = 2$, impossible.

$\boxed{\text{Therefore, the equation has no solutions.}}$

9. $13^{x^3-x+1} = 13$

We have $13^{x^3-x+1} = 13$. Then

$$x^3 - x + 1 = 1$$
$$x^3 - x = 0$$
$$x(x^2 - 1) = 0$$
$$x(x-1)(x+1) = 0.$$

It follows that $\begin{bmatrix} x = 0 \\ x - 1 = 0 \\ x + 1 = 0 \end{bmatrix}$. We obtain $\begin{bmatrix} x = 0 \\ x = 1 \\ x = -1 \end{bmatrix}$.

$\boxed{\text{Therefore, } x \in \{-1, 0, 1\}.}$

10. $15^{x^3-3x^2+1} = 15^{-1}$

We have $15^{x^3-3x^2+1} = 15^{-1}$. Then

$$x^3 - 3x^2 + 1 = -1$$
$$x^3 - 3x^2 + 2 = 0$$
$$x^3 - x^2 - 2x^2 + 2 = 0$$
$$x^2(x-1) - 2(x^2 - 1) = 0$$
$$x^2(x-1) - 2(x-1)(x+1) = 0$$
$$(x-1)(x^2 - 2x - 2) . = 0$$

It follows that $\begin{bmatrix} x - 1 = 0 \\ x^2 - 2x - 2 = 0 \end{bmatrix}$.

- If $x - 1 = 0$, we obtain $x = 1$.
- If $x^2 - 2x - 2 = 0$, we obtain $\Delta' = (b')^2 - ac$
$$= (-1)^2 - (1)(-2)$$
$$= 1 + 2$$
$$= 3.$$

It follows that
$$x_1 = \frac{-b' + \sqrt{\Delta'}}{a} = \frac{1 + \sqrt{3}}{1} = 1 + \sqrt{3}$$
and
$$x_2 = \frac{-b' - \sqrt{\Delta'}}{a} = \frac{1 - \sqrt{3}}{1} = 1 - \sqrt{3}.$$

Therefore, $x \in \left\{1 - \sqrt{3}, 1, 1 + \sqrt{3}\right\}$.

Problem 13. Solve the following exponential equations:

1. $2^x = 16$
2. $6^x = 216$
3. $3^x = 81$
4. $0.4^x = \dfrac{25}{4}$
5. $3^x = \dfrac{9}{\sqrt[3]{9}}$
6. $0.2^x = \sqrt[3]{0.008}$
7. $\left(\dfrac{1}{2}\right)^x = \dfrac{1}{128}$
8. $\left(\dfrac{1}{4}\right)^x = \dfrac{1}{256}$
9. $3^{-x} = \dfrac{1}{243}$
10. $\left(\dfrac{2}{3}\right)^x = 1.5$
11. $\left(\dfrac{2}{7}\right)^x = \dfrac{343}{8}$
12. $\left(\dfrac{3}{4}\right)^x = \dfrac{64}{27}$
13. $25^x = 5$

14. $3^{2^x} = 6561$

15. $81^{4^x} = 9$

16. $2^x \times 5^x = 1$

17. $\left(2-\sqrt{3}\right)^x + \left(2+\sqrt{3}\right)^x = 2$

18. $3 \times 4^x + 2 \times 9^x - 5 \times 6^x = 0$

19. $\left(\sqrt{2+\sqrt{3}}\right)^x + \left(\sqrt{2-\sqrt{3}}\right)^x = 4$

Solution. Solve the following exponential equations:

1. $2^x = 16$
   We have
   $$2^x = 16$$
   $$2^x = 2^4.$$

   Therefore, $x = 4.$

2. $6^x = 216$
   We have
   $$6^x = 216$$
   $$6^x = 6^3.$$

   Therefore, $x = 3.$

3. $3^x = 81$
   We have
   $$3^x = 81$$
   $$3^x = 3^4.$$

   Therefore, $x = 4.$

4. $0.4^x = \dfrac{25}{4}$
   We have
   $$0.4^x = \dfrac{25}{4}$$

$$\left(\frac{4}{10}\right)^x = \frac{25}{4}$$
$$\left(\frac{2}{5}\right)^x = \left(\frac{5}{2}\right)^2$$
$$\left(\frac{2}{5}\right)^x = \left(\frac{2}{5}\right)^{-2}.$$

Therefore, $x = -2$.

5. $3^x = \dfrac{9}{\sqrt[3]{9}}$
We have
$$3^x = \frac{9}{\sqrt[3]{9}}$$
$$3^x = \frac{3^2}{\sqrt[3]{3^2}}$$
$$3^x = \frac{3^2}{3^{\frac{2}{3}}}$$
$$3^x = 3^{2-\frac{2}{3}}$$
$$3^x = 3^{\frac{4}{3}}.$$

Therefore, $x = \dfrac{4}{3}$.

6. $0.2^x = \sqrt[3]{0.008}$
We have
$$0.2^x = \sqrt[3]{0.008}$$
$$0.2^x = \sqrt[3]{(0.2)^3}$$
$$0.2^x = 0.2.$$

Therefore, $x = 1$.

7. $\left(\dfrac{1}{2}\right)^x = \dfrac{1}{128}$
We have
$$\left(\frac{1}{2}\right)^x = \frac{1}{128}$$

$$\left(\frac{1}{2}\right)^x = \frac{1}{2^7}$$
$$\left(\frac{1}{2}\right)^x = \left(\frac{1}{2}\right)^7.$$

Therefore, $x = 7$.

8. $\left(\dfrac{1}{4}\right)^x = \dfrac{1}{256}$
We have
$$\left(\frac{1}{4}\right)^x = \frac{1}{256}$$
$$\left(\frac{1}{4}\right)^x = \frac{1}{4^4}$$
$$\left(\frac{1}{4}\right)^x = \left(\frac{1}{4}\right)^4.$$

Therefore, $x = 4$.

9. $3^{-x} = \dfrac{1}{243}$
We have
$$3^{-x} = \frac{1}{243}$$
$$3^{-x} = \frac{1}{3^5}$$
$$\frac{1}{3^x} = \frac{1}{3^5}$$
$$\left(\frac{1}{3}\right)^x = \left(\frac{1}{3}\right)^5.$$

Therefore, $x = 5$.

10. $\left(\dfrac{2}{3}\right)^x = 1.5$
We have
$$\left(\frac{2}{3}\right)^x = 1.5$$

$$\left(\frac{2}{3}\right)^x = \frac{3}{2}$$
$$\left(\frac{2}{3}\right)^x = \left(\frac{2}{3}\right)^{-1}.$$

Therefore, $x = -1$.

11. $\left(\frac{2}{7}\right)^x = \frac{343}{8}$

We have
$$\left(\frac{2}{7}\right)^x = \frac{343}{8}$$
$$\left(\frac{2}{7}\right)^x = \frac{7^3}{2^3}$$
$$\left(\frac{2}{7}\right)^x = \left(\frac{7}{2}\right)^3$$
$$\left(\frac{2}{7}\right)^x = \left(\frac{2}{7}\right)^{-3}.$$

Therefore, $x = -3$.

12. $\left(\frac{3}{4}\right)^x = \frac{64}{27}$

We have
$$\left(\frac{3}{4}\right)^x = \frac{64}{27}$$
$$\left(\frac{3}{4}\right)^x = \frac{4^3}{3^3}$$
$$\left(\frac{3}{4}\right)^x = \left(\frac{4}{3}\right)^3$$
$$\left(\frac{3}{4}\right)^x = \left(\frac{3}{4}\right)^{-3}.$$

Therefore, $x = -3$.

13. $25^x = 5$

We have
$$25^x = 5$$

$$\left(5^2\right)^x = 5$$
$$5^{2x} = 5.$$

Therefore, $2x = 1$ or $x = \dfrac{1}{2}$.

14. $3^{2^x} = 6561$
    We have $3^{2^x} = 6561$ or $3^{2^x} = 3^8$.
    Then $2^x = 8 = 2^3$.
    Therefore, $x = 3$.

15. $81^{4^x} = 9$
    We have
    $$81^{4^x} = 9$$
    $$\left(9^2\right)^{4^x} = 9$$
    $$9^{2 \times 4^x} = 9.$$

    It follows that
    $$2 \times 4^x = 1$$
    $$2 \times \left(2^2\right)^x = 1$$
    $$2^{2x+1} = 1 = 2^0.$$

    Hence, $2x + 1 = 0$.
    Therefore, $x = -\dfrac{1}{2}$.

16. $2^x \times 5^x = 1$
    We have
    $$2^x \times 5^x = 1$$
    $$(2 \times 5)^x = 1$$
    $$10^x = 10^0.$$

    Therefore, $x = 0$.

17. $\left(2 - \sqrt{3}\right)^x + \left(2 + \sqrt{3}\right)^x = 2$
    We have $\left(2 - \sqrt{3}\right)\left(2 + \sqrt{3}\right) = 2^2 - \sqrt{3^2} = 4 - 3 = 1.$

Then $2 + \sqrt{3} = \dfrac{1}{2 - \sqrt{3}}$.

The given equation can be written as

$$\left(2 - \sqrt{3}\right)^x + \left(\dfrac{1}{2 - \sqrt{3}}\right)^x = 2 \text{ or } \left(2 - \sqrt{3}\right)^x + \dfrac{1}{\left(2 - \sqrt{3}\right)^x} - 2 = 0.$$

Let $t = \left(2 - \sqrt{3}\right)^x$. We obtain $t > 0$ for all $x \in \mathbb{R}$.

Then $t + \dfrac{1}{t} - 2 = 0$ or $t^2 - 2t + 1 = 0$.

It implies that $(t - 1)^2 = 0$.

As a result, we obtain $t - 1 = 0$ or $t = 1$.

Hence, $\left(2 - \sqrt{3}\right)^x = 1 = \left(2 - \sqrt{3}\right)^0$.

$\boxed{\text{Therefore, } x = 0.}$

18. $3 \times 4^x + 2 \times 9^x - 5 \times 6^x = 0$

Divide both sides of the given equation by 9^x, we obtain

$$\dfrac{3 \times 4^x}{9^x} + \dfrac{2 \times 9^x}{9^x} - \dfrac{5 \times 6^x}{9^x} = 0$$

$$3 \left(\dfrac{4}{9}\right)^x + 2 - 5 \left(\dfrac{6}{9}\right)^x = 0$$

$$3 \left[\left(\dfrac{2}{3}\right)^2\right]^x - 5 \left(\dfrac{2}{3}\right)^x + 2 = 0$$

$$3 \left[\left(\dfrac{2}{3}\right)^x\right]^2 - 5 \left(\dfrac{2}{3}\right)^x + 2 = 0.$$

Let $t = \left(\dfrac{2}{3}\right)^x$. We obtain $t > 0$ for all $x \in \mathbb{R}$.

Then $3t^2 - 5t + 2 = 0$.

Since the sum of all coefficients of the last quadratic equation is zero, we obtain

$$t = 1 \text{ or } t = \dfrac{2}{3}.$$

- If $t = 1$, we obtain $\left(\dfrac{2}{3}\right)^x = 1 = \left(\dfrac{2}{3}\right)^0$. Then $x = 0$.

- If $t = \dfrac{2}{3}$, we obtain $\left(\dfrac{2}{3}\right)^x = \dfrac{2}{3}$. Then $x = 1$.

Therefore, $x \in \{0, 1\}$.

19. $\left(\sqrt{2+\sqrt{3}}\right)^x + \left(\sqrt{2-\sqrt{3}}\right)^x = 4$

    Let $a = \left(\sqrt{2+\sqrt{3}}\right)^x$ and $b = \left(\sqrt{2-\sqrt{3}}\right)^x$. It follows that

    $$ab = \left(\sqrt{2+\sqrt{3}}\right)^x \left(\sqrt{2-\sqrt{3}}\right)^x$$
    $$= \left[\sqrt{\left(2+\sqrt{3}\right)\left(2-\sqrt{3}\right)}\right]^x$$
    $$= \left(\sqrt{2^2 - \sqrt{3}^2}\right)^x$$
    $$= \left(\sqrt{4-3}\right)^x = 1^x.$$

Then
$$ab = 1. \tag{1}$$

The given equation can be written as
$$a + b = 4. \tag{2}$$

From (1) and (2), it turns out that a and b are the roots of the quadratic equation $X^2 - 4X + 1 = 0$.

The discriminant of the last quadratic equation is
$$\Delta' = (b')^2 - ac = (-2)^2 - (1)(1) = 4 - 1 = 3.$$

Then $X = \dfrac{-b' \pm \sqrt{\Delta'}}{a} = 2 \pm \sqrt{3}$.

We obtain $\begin{cases} a = 2 - \sqrt{3} \\ b = 2 + \sqrt{3} \end{cases}$ or $\begin{cases} a = 2 + \sqrt{3} \\ b = 2 - \sqrt{3} \end{cases}$.

- If $\begin{cases} a = 2 - \sqrt{3} \\ b = 2 + \sqrt{3} \end{cases}$, we obtain $\begin{cases} \left(\sqrt{2+\sqrt{3}}\right)^x = 2 - \sqrt{3} \\ \left(\sqrt{2-\sqrt{3}}\right)^x = 2 + \sqrt{3} \end{cases}$.

    Then
    $$\begin{cases} \left(\sqrt{2+\sqrt{3}}\right)^x = \dfrac{1}{2+\sqrt{3}} \\ \left(\sqrt{2-\sqrt{3}}\right)^x = \dfrac{1}{2-\sqrt{3}} \end{cases}.$$

We obtain

$$\begin{cases} \left(\sqrt{2+\sqrt{3}}\right)^x = \dfrac{1}{\sqrt{(2+\sqrt{3})^2}} \\ \left(\sqrt{2-\sqrt{3}}\right)^x = \dfrac{1}{\sqrt{(2-\sqrt{3})^2}} \end{cases}.$$

It implies that

$$\begin{cases} \left(\sqrt{2+\sqrt{3}}\right)^x = \left(\sqrt{2+\sqrt{3}}\right)^{-2} \\ \left(\sqrt{2-\sqrt{3}}\right)^x = \left(\sqrt{2-\sqrt{3}}\right)^{-2} \end{cases}.$$

It follows that $x = -2$.

- If $\begin{cases} a = 2+\sqrt{3} \\ b = 2-\sqrt{3} \end{cases}$, we obtain

$$\begin{cases} \sqrt{\left(2+\sqrt{3}\right)^x} = 2+\sqrt{3} = \sqrt{\left(2+\sqrt{3}\right)^2} \\ \sqrt{\left(2-\sqrt{3}\right)^x} = 2-\sqrt{3} = \sqrt{\left(2-\sqrt{3}\right)^2} \end{cases}.$$

Then $x = 2$.

Therefore, $x \in \{-2, 2\}$.

Problem 14. Solve the following exponential equations:

1. $3^{2x} = 27$
2. $2^{5x} = \dfrac{1}{128}$
3. $0.5^{-2x} = 0.125$
4. $17^{3x} = 1$
5. $0.27^x = 0.04$
6. $3^{x+1} = 9$
7. $8^{x-3} = 16$
8. $32^x = 16^{1-x}$
9. $9^{4x-2} = \dfrac{1}{81}$
10. $0.5^{3x-2} = 0.125$
11. $\sqrt{7^{3x+1}} = 49$
12. $4^x \times 16^{3x-1} = 8$
13. $8^{2x} \times 4^{2x-1} = 16$
14. $2^x \times 3^x = 216$

15. $5^x \times 2^x = 0.001$

16. $5^{x-1} \times 2^{x-1} = 0.001$

17. $7^{1-x} \times 4^{1-x} = \dfrac{1}{28}$

18. $\sqrt{12^x} \times \sqrt{3^x} = \dfrac{1}{6}$

19. $\sqrt{7^{x-1}} \times \sqrt{2^{x-1}} = 1$

20. $2^x + 2^{x+2} = 20$

21. $3^x + 3^{x+1} = 12$

22. $5^x + 5^{x-1} = 30$

23. $4^{x-1} + 4^x + 4^{x+1} = 84$

Solution. Solve the following exponential equations:

1. $3^{2x} = 27$
   We have
$$3^{2x} = 27$$
$$3^{2x} = 3^3.$$
   Then $2x = 3$. It follows that $x = \dfrac{3}{2}$.

   $\boxed{\text{Hence, } x = \dfrac{3}{2}.}$

2. $2^{5x} = \dfrac{1}{128}$
   We have
$$2^{5x} = \dfrac{1}{128}$$
$$2^{5x} = \dfrac{1}{2^7}$$
$$2^{5x} = 2^{-7}.$$

   $\boxed{\text{Therefore, } 5x = -7 \text{ or } x = -\dfrac{7}{5}.}$

3. $0.5^{-2x} = 0.125$
   We have
$$0.5^{-2x} = 0.125$$
$$0.5^{-2x} = 0.5^3.$$
   Then $-2x = 3$. It follows that $x = -\dfrac{3}{2}$.

   $\boxed{\text{Therefore, } x = -\dfrac{3}{2}.}$

4. $17^{3x} = 1$

We have $17^{3x} = 1 = 17^0$. Then $3x = 0$. It follows that $x = \dfrac{0}{3} = 0$.

$\boxed{\text{Therefore, } x = 0.}$

5. $0.2^{7x} = 0.04$

We have
$$0.2^{7x} = 0.04$$
$$0.2^{7x} = 0.2^2.$$

Then $7x = 2$. It implies that $x = \dfrac{2}{7}$.

$\boxed{\text{Therefore, } x = \dfrac{2}{7}.}$

6. $3^{x+1} = 9$

We have
$$3^{x+1} = 9$$
$$3^{x+1} = 3^2.$$

Then $x + 1 = 2$. It follows that $x = 2 - 1 = 1$.

$\boxed{\text{Therefore, } x = 1.}$

7. $8^{x-3} = 16$

We have
$$8^{x-3} = 16$$
$$\left(2^3\right)^{x-3} = 2^4$$
$$2^{3(x-3)} = 2^4$$
$$2^{3x-9} = 2^4.$$

Then $3x - 9 = 4$. It implies that $3x = 4 + 9 = 13$.

$\boxed{\text{Therefore, } x = \dfrac{13}{3}.}$

8. $32^x = 16^{1-x}$

We have
$$32^x = 16^{1-x}$$

Chapter 5. Solutions

$$(2^5)^x = (2^3)^{1-x}$$
$$2^{5x} = 2^{3(1-x)}$$
$$2^{5x} = 2^{3-3x}.$$

Then $5x = 3-3x$. It follows that $5x+3x = 3$. Hence, $8x = 3$.

$\boxed{\text{Therefore, } x = \dfrac{3}{8}.}$

9. $9^{4x-2} = \dfrac{1}{81}$

We have

$$9^{4x-2} = \frac{1}{81}$$
$$9^{4x-2} = \frac{1}{9^2}$$
$$9^{4x-2} = 9^{-2}.$$

Then $4x - 2 = -2$. It follows that $4x = -2 + 2$. It turns out that $4x = 0$. Hence, $x = 0$.

$\boxed{\text{Therefore, } x = 0.}$

10. $0.5^{3x-2} = 0.125$

We have

$$0.5^{3x-2} = 0.125$$
$$0.5^{3x-2} = 0.5^3.$$

Then $3x - 2 = 3$. It follows that $3x = 2 + 3 = 5$.

$\boxed{\text{Therefore, } x = \dfrac{5}{3}.}$

11. $\sqrt{7^{3x+1}} = 49$

The equation is well-defined if and only if $3x + 1 \geq 0$.
It follows that $x \geq -\dfrac{1}{3}$.
We have

$$\sqrt{7^{3x+1}} = 49$$
$$\sqrt{7^{3x+1}} = 7^2$$
$$7^{3x+1} = (7^2)^2$$

120

$$7^{3x+1} = 7^4.$$

Then $3x + 1 = 4$. It follows that $3x = 4 - 1 = 3$.

Therefore, $x = \dfrac{3}{3} = 1$.

12. $4^x \times 16^{3x-1} = 8$

We have
$$4^x \times 16^{3x-1} = 8$$
$$2^{2x} \times \left(2^4\right)^{3x-1} = 2^3$$
$$2^{2x} \times 2^{4(3x-1)} = 2^3$$
$$2^{2x} \times 2^{12x-4} = 2^3$$
$$2^{2x+12x-4} = 2^3$$
$$2^{14x-4} = 2^3.$$

Then $14x - 4 = 3$. It follows that $14x = 3 + 4 = 7$.

Therefore, $x = \dfrac{7}{14} = \dfrac{1}{2}$.

13. $8^{2x} \times 4^{2x-1} = 16$

We have
$$8^{2x} \times 4^{2x-1} = 16$$
$$\left(2^3\right)^{2x} \times \left(2^2\right)^{2x-1} = 2^4$$
$$2^{6x} \times 2^{2(2x-1)} = 2^4$$
$$2^{6x+2(2x-1)} = 2^4$$
$$2^{6x+4x-2} = 2^4$$
$$2^{10x-2} = 2^4.$$

It follows that $10x - 2 = 4$. Then $10x = 6$.

Hence, $x = \dfrac{6}{10} = \dfrac{3}{5}$.

Therefore, $x = \dfrac{3}{5}$.

14. $2^x \times 3^x = 216$

We have
$$2^x \times 3^x = 216$$

$$(2 \times 3)^x = 6^3$$
$$6^x = 6^3.$$

$\boxed{\text{Therefore, } x = 3.}$

15. $5^x \times 2^x = 0.001$
We have
$$5^x \times 2^x = 0.001$$
$$(5 \times 2)^x = 10^{-3}$$
$$10^x = 10^{-3}.$$

$\boxed{\text{Therefore, } x = -3.}$

16. $5^{x-1} \times 2^{x-1} = 0.001$
We have
$$5^{x-1} \times 2^{x-1} = 0.001$$
$$(5 \times 2)^{x-1} = 10^{-3}$$
$$10^{x-1} = 10^{-3}.$$

Then $x - 1 = -3$. It follows that $x = -3 + 1 = -2$.
$\boxed{\text{Therefore, } x = -2.}$

17. $7^{1-x} \times 4^{1-x} = \dfrac{1}{28}$
We have
$$7^{1-x} \times 4^{1-x} = \dfrac{1}{28}$$
$$(7 \times 4)^{1-x} = 28^{-1}$$
$$28^{1-x} = 28^{-1}.$$

Then $1 - x = -1$. It follows that $x = 1 + 1 = 2$.
$\boxed{\text{Therefore, } x = 2.}$

18. $\sqrt{12^x} \times \sqrt{3^x} = \dfrac{1}{6}$
We have
$$\sqrt{12^x} \times \sqrt{3^x} = \dfrac{1}{6}$$

$$\sqrt{12^x \times 3^x} = \frac{1}{6}$$
$$\sqrt{36^x} = 6^{-1}$$
$$36^x = 6^{-2}$$
$$6^{2x} = 6^{-2}.$$

Then $2x = -2$. It follows that $x = -\frac{2}{2} = -1$.

Therefore, $x = -1.$

19. $\sqrt{7^{x-1}} \times \sqrt{2^{x-1}} = 1$
We have
$$\sqrt{7^{x-1}} \times \sqrt{2^{x-1}} = 1$$
$$\sqrt{7^{x-1} \times 2^{x-1}} = 1$$
$$\sqrt{14^{x-1}} = 1$$
$$14^{x-1} = 1 = 14^0.$$

Then $x - 1 = 0$. It follows that $x = 1$.
Therefore, $x = 1.$

20. $2^x + 2^{x+2} = 20$
We have
$$2^x + 2^{x+2} = 20$$
$$2^x + 2^2 \times 2^x = 20$$
$$2^x + 4 \times 2^x = 20$$
$$5 \times 2^x = 20$$
$$2^x = \frac{20}{5} = 4$$
$$2^x = 2^2.$$

Therefore, $x = 2.$

21. $3^x + 3^{x+1} = 12$
We have
$$3^x + 3^{x+1} = 12$$

$$3^x + 3 \times 3^x = 12$$
$$4 \times 3^x = 12$$
$$3^x = \frac{12}{4} = 3.$$

It follows that $x = 1$.
Therefore, $x = 1$.

22. $5^x + 5^{x-1} = 30$
We have

$$5^x + 5^{x-1} = 30$$
$$5 \times 5^{x-1} + 5^{x-1} = 30$$
$$6 \times 5^{x-1} = 30$$
$$5^{x-1} = \frac{30}{6}$$
$$5^{x-1} = 5.$$

Then $x - 1 = 1$. It follows that $x = 1 + 1 = 2$.
Therefore, $x = 2$.

23. $4^{x-1} + 4^x + 4^{x+1} = 84$
We have

$$4^{x-1} + 4 \times 4^{x-1} + 4^2 \times 4^{x-1} = 84$$
$$4^{x-1} + 4 \times 4^{x-1} + 16 \times 4^{x-1} = 84$$
$$21 \times 4^{x-1} = 84$$
$$4^{x-1} = \frac{84}{21} = 4.$$

Then $x - 1 = 1$. It follows that $x = 1 + 1 = 2$.
Therefore, $x = 2$.

Problem 15. Solve the following exponential equations:

1. $3^{x^2} = 81^{x-1}$

2. $9^{(2x-1)^2} = 3^{x+3}$

3. $2^{2x+3} - 2^x = 1 - 2^{x+3}$

4. $3^{x^2+4x} = \dfrac{1}{27}$

5. $3^{5x} \times 9^{x^2} = 27$

6. $4^{3x^2+2x+1} = 16$

Solution. Solve the following exponential equations:

1. $3^{x^2} = 81^{x-1}$

   We have
   $$3^{x^2} = 81^{x-1}$$
   $$3^{x^2} = \left(3^4\right)^{x-1}$$
   $$3^{x^2} = 3^{4(x-1)}.$$

   Then $x^2 = 4(x-1)$. It follows that $x^2 - 4(x-1) = 0$. We obtain $x^2 - 4x + 4 = 0$. Then $(x-2)^2 = 0$. Hence, $x - 2 = 0$. It implies that $x = 2$.

   $\boxed{\text{Therefore, } x = 2.}$

2. $9^{(2x-1)^2} = 3^{x+3}$

   We have
   $$9^{(2x-1)^2} = 3^{x+3}$$
   $$\left(3^2\right)^{(2x-1)^2} = 3^{x+3}$$
   $$3^{2(2x-1)^2} = 3^{x+3}.$$

   Then $2(2x-1)^2 = x+3$ or $2(4x^2 - 4x + 1) = x + 3$. It follows that $8x^2 - 8x + 2 - x - 3 = 0$. We obtain $8x^2 - 9x - 1 = 0$.
   The discriminant of the last quadratic equation is
   $$\Delta = b^2 - 4ac$$
   $$= (-9)^2 - 4(8)(-1)$$
   $$= 81 + 32 = 113.$$

   We obtain
   $$x_1 = \frac{-b + \sqrt{\Delta}}{2a} = \frac{-(-9) + \sqrt{113}}{2(8)} = \frac{9 + \sqrt{113}}{16}$$
   and
   $$x_2 = \frac{-b - \sqrt{\Delta}}{2a} = \frac{-(-9) - \sqrt{113}}{2(8)} = \frac{9 - \sqrt{113}}{16}.$$

   $\boxed{\text{Therefore, } x \in \left\{\dfrac{9 + \sqrt{113}}{16}, \dfrac{9 - \sqrt{113}}{16}\right\}.}$

3. $2^{2x+3} - 2^x = 1 - 2^{x+3}$

We have
$$2^{2x+3} - 2^x = 1 - 2^{x+3}$$
$$2^3 \times 2^{2x} - 2^x = 1 - 2^3 \times 2^x$$
$$8 \times (2^x)^2 - 2^x = 1 - 8 \times 2^x$$
$$8 \times (2^x)^2 - 2^x + 8 \times 2^x - 1 = 0$$
$$8 \times (2^x)^2 + 7 \times 2^x - 1 = 0.$$

Let $t = 2^x$. Then $t > 0$ for all $x \in \mathbb{R}$. The given equation can be written as $8t^2 + 7t - 1 = 0$.
Then $(t+1)(8t-1) = 0$. It follows that $8t - 1 = 0$.
Then $t = \dfrac{1}{8}$.
In this case, we obtain $2^x = \dfrac{1}{8} = \dfrac{1}{2^3} = 2^{-3}$.

Therefore, $x = -3$.

4. $3^{x^2+4x} = \dfrac{1}{27}$

We have
$$3^{x^2+4x} = \dfrac{1}{27}$$
$$3^{x^2+4x} = \dfrac{1}{3^3}$$
$$3^{x^2+4x} = 3^{-3}.$$

Then $x^2 + 4x = -3$. It follows that $x^2 + 4x + 3 = 0$.
We obtain $(x+1)(x+3) = 0$. Then $\begin{bmatrix} x+1 = 0 \\ x+3 = 0 \end{bmatrix}$ or $\begin{bmatrix} x = -1 \\ x = -3 \end{bmatrix}$.

Therefore, $x \in \{-3, -1\}$.

5. $3^{5x} \times 9^{x^2} = 27$

We have
$$3^{5x} \times 9^{x^2} = 27$$
$$3^{5x} \times (3^2)^{x^2} = 3^3$$
$$3^{5x} \times 3^{2x^2} = 3^3$$

$$3^{5x+2x^2} = 3^3.$$

Then $5x + 2x^2 = 3$. It follows that $2x^2 + 5x - 3 = 0$.
We obtain $(x+3)(2x-1) = 0$.

Hence, $\begin{bmatrix} x+3=0 \\ 2x-1=0 \end{bmatrix}$. Then $\begin{bmatrix} x = -3 \\ x = \dfrac{1}{2} \end{bmatrix}$.

Therefore, $x \in \left\{-3, \dfrac{1}{2}\right\}$.

6. $4^{3x^2+2x+1} = 16$
We have
$$4^{3x^2+2x+1} = 16$$
$$4^{3x^2+2x+1} = 4^2.$$

Then $3x^2 + 2x + 1 = 2$. It follows that $3x^2 + 2x - 1 = 0$.
We obtain $(3x-1)(x+1) = 0$.

As a result, $\begin{bmatrix} 3x-1=0 \\ x+1=0 \end{bmatrix}$ or $\begin{bmatrix} x = \dfrac{1}{3} \\ x = -1 \end{bmatrix}$.

Therefore, $x \in \left\{-1, \dfrac{1}{3}\right\}$.

Problem 16. Solve the following exponential equations:

1. $9^x + 3^x - 2 = 0$
2. $16^x - 5 \times 4^x + 4 = 0$
3. $4^x - 3 \times 2^x + 2 = 0$
4. $64^x - 3 \times 2^{3x+1} + 8 = 0$
5. $9^x - 4 \times 3^{x+1} + 27 = 0$

Solution. Solve the following exponential equations:

1. $9^x + 3^x - 2 = 0$
We have
$$9^x + 3^x - 2 = 0$$
$$\left(3^2\right)^x + 3^x - 2 = 0$$
$$(3^x)^2 + 3^x - 2 = 0.$$

Let $t = 3^x$. Then $t > 0$ for all $x \in \mathbb{R}$.
The given equation can be written as $t^2 + t - 2 = 0$.
Then $(t-1)(t+2) = 0$. It follows that $t-1 = 0$ since $t+2 > 0$.
Then $t = 1$ since $t + 2 > 0$.
For $t = 1$, we obtain $3^x = 1 = 3^0$. Then $x = 0$.
$\boxed{\text{Therefore, } x = 0.}$

2. $16^x - 5 \times 4^x + 4 = 0$

$$16^x - 5 \times 4^x + 4 = 0$$
$$\left(4^2\right)^x - 5 \times 4^x + 4 = 0$$
$$\left(4^x\right)^2 - 5 \times 4^x + 4 = 0$$

Let $t = 4^x$. Then $t > 0$ for all $x \in \mathbb{R}$.
The given equation can be written as $t^2 - 5t + 4 = 0$.
Then $(t-1)(t-4) = 0$.
It follows that $\left[\begin{array}{l} t - 1 = 0 \\ t - 4 = 0 \end{array} \right.$ or $\left[\begin{array}{l} t = 1 \\ t = 4 \end{array} \right.$.

- If $t = 1$, we obtain $4^x = 1 = 4^0$. Then $x = 0$.
- If $t = 4$, we obtain $4^x = 4$. Then $x = 1$.

$\boxed{\text{Therefore, } x \in \{0, 1\}.}$

3. $4^x - 3 \times 2^x + 2 = 0$
We have

$$4^x - 3 \times 2^x + 2 = 0$$
$$\left(2^2\right)^x - 3 \times 2^x + 2 = 0$$
$$\left(2^x\right)^2 - 3 \times 2^x + 2 = 0.$$

Let $t = 2^x$. Then $t > 0$ for all $x \in \mathbb{R}$.
The given equation can be written as $t^2 - 3t + 2 = 0$.
It follows that $(t-1)(t-2) = 0$.
Then $\left[\begin{array}{l} t - 1 = 0 \\ t - 2 = 0 \end{array} \right.$ or $\left[\begin{array}{l} t = 1 \\ t = 2 \end{array} \right.$.

- If $t = 1$, then $2^x = 1 = 2^0$. It follows that $x = 0$.
- If $t = 2$, then $2^x = 2$. It implies that $x = 1$.

$\boxed{\text{Therefore, } x \in \{0, 1\}.}$

4. $64^x - 3 \times 2^{3x+1} + 8 = 0$
We have
$$64^x - 3 \times 2^{3x+1} + 8 = 0$$
$$(8^2)^x - 3(2) 2^{3x} + 8 = 0$$
$$(8^x)^2 - 6(8^x) + 8 = 0.$$

Let $t = 8^x$. Then $t > 0$ for all $x \in \mathbb{R}$.
The given equation can be written as $t^2 - 6t + 8 = 0$.
Let $(t-2)(t-4) = 0$. Then $\begin{bmatrix} t - 2 = 0 \\ t - 4 = 0 \end{bmatrix}$ or $\begin{bmatrix} t = 2 \\ t = 4 \end{bmatrix}$.

- If $t = 2$, we obtain $8^x = 2$. Then $2^{3x} = 2$. It follows that $3x = 1$. As a result, $x = \dfrac{1}{3}$.
- If $t = 4$, then $8^x = 4$. It follows that $2^{3x} = 2^2$. We obtain $3x = 2$. Hence, $x = \dfrac{2}{3}$.

Therefore, $x \in \left\{ \dfrac{1}{3}, \dfrac{2}{3} \right\}$.

5. $9^x - 4 \times 3^{x+1} + 27 = 0$
We have
$$9^x - 4 \times 3^{x+1} + 27 = 0$$
$$(3^2)^x - 4(3) 3^x + 27 = 0$$
$$(3^x)^2 - 12 \times 3^x + 27 = 0.$$

Let $t = 3^x$. Then $t > 0$ for all $x \in \mathbb{R}$.
The given equation can be written as $t^2 - 12t + 27 = 0$.
Then $(t-3)(t-9) = 0$. We obtain $\begin{bmatrix} t - 3 = 0 \\ t - 9 = 0 \end{bmatrix}$ or $\begin{bmatrix} t = 3 \\ t = 9 \end{bmatrix}$.

- If $t = 3$, then $3^x = 3$. It follows that $x = 1$.
- If $t = 9$, then $3^x = 9 = 3^2$. We obtain $x = 2$.

Therefore, $x \in \{1, 2\}$.

Problem 17. Solve the following exponential equations:

1. $2^{x^2 - 2x + 1} = 1$

2. $5^{2x+1} - 3 \times 5^{2x-1} = 550$

3. $3^{2x-1} \times 5^{3x+2} = \dfrac{9}{5} \times 5^{2x} \times 3^{3x}$

4. $4^x - 3^{x-\frac{1}{2}} = 3^{x+\frac{1}{2}} - 2^{2x-1}$

5. $3 \times 16^x + 37 \times 36^x = 26 \times 81^x$

6. $3^{2x^2+6x-9} + 4 \times 15^{x^2+3x-5} = 3 \times 5^{2x^2+6x-9}$

7. $27^x + 12^x = 2 \times 8^x$

8. $5 \times 2^{3x-3} - 3 \times 2^{5-3x} + 7 = 0$

9. $3^{2x^2-6x+3} + 6^{x^2-3x+1} = 2^{2x^2-6x+3}$

Solution. Solve the following exponential equations:

1. $2^{x^2-2x+1} = 1$
   We have $2^{x^2-2x+1} = 1$. Then $2^{(x-1)^2} = 2^0$.
   It follows that $(x-1)^2 = 0$. Then $x - 1 = 0$.
   $\boxed{\text{Therefore, } x = 1.}$

2. $5^{2x+1} - 3 \times 5^{2x-1} = 550$
   We have
   $$5^{2x+1} - 3 \times 5^{2x-1} = 550$$
   $$5^2 \times 5^{2x-1} - 3 \times 5^{2x-1} = 550$$
   $$25 \times 5^{2x-1} - 3 \times 5^{2x-1} = 550$$
   $$22 \times 5^{2x-1} = 550$$
   $$5^{2x-1} = 25$$
   $$5^{2x-1} = 5^2.$$
   Then $2x - 1 = 2$. It follows that $2x = 3$.
   $\boxed{\text{Therefore, } x = \dfrac{3}{2}.}$

3. $3^{2x-1} \times 5^{3x+2} = \dfrac{9}{5} \times 5^{2x} \times 3^{3x}$
   We have $3^{2x-1} \times 5^{3x+2} = \dfrac{9}{5} \times 5^{2x} \times 3^{3x}$.
   Then
   $$\dfrac{3^{2x-1} \times 5^{3x+2} \times 5}{3^{3x} \times 5^{2x} \times 9} = 1$$

or
$$3^{2x-1-3x-2} \times 5^{3x+2+1-2x} = 1.$$

It implies that $3^{-x-3} \times 5^{x+3} = 1$. Then $\dfrac{5^{x+3}}{3^{x+3}} = 1$.

We obtain $\left(\dfrac{5}{3}\right)^{x+3} = 1$.

As a result, $x + 1 = 0$. Hence, $x = -1$.

$\boxed{\text{Therefore, } x = -1.}$

4. $4^x - 3^{x-\frac{1}{2}} = 3^{x+\frac{1}{2}} - 2^{2x-1}$

We have
$$4^x - 3^{x-\frac{1}{2}} = 3^{x+\frac{1}{2}} - 2^{2x-1}$$
$$2 \times 2^{2x-1} - 3^{x-\frac{1}{2}} = 3 \times 3^{x-\frac{1}{2}} - 2^{2x-1}$$
$$2 \times 2^{2x-1} + 2^{2x-1} = 3 \times 3^{x-\frac{1}{2}} + 3^{x-\frac{1}{2}}$$
$$3 \times 2^{2x-1} = 4 \times 3^{x-\frac{1}{2}}.$$

Then $\dfrac{2^{2x-1} \times 3}{4 \times 3^{x-\frac{1}{2}}} = 1$. It follows that $\dfrac{2^{2x-1-2}}{3^{x-\frac{1}{2}-1}} = 1$.

Then $\dfrac{2^{2x-3}}{3^{x-\frac{3}{2}}} = 1$. We obtain $\dfrac{2^{2x-3}}{3^{\frac{2x-3}{2}}} = 1$.

Hence, $\left(\dfrac{2}{3^{\frac{1}{2}}}\right)^{2x-3} = \left(\dfrac{2}{3^{\frac{1}{2}}}\right)^0$.

Then $2x - 3 = 0$. As a result, $x = \dfrac{3}{2}$.

$\boxed{\text{Therefore, } x = \dfrac{3}{2}.}$

5. $3 \times 16^x + 37 \times 36^x = 26 \times 81^x$

We have
$$3 \times 16^x + 37 \times 36^x = 26 \times 81^x$$
$$3 \times \left(4^2\right)^x + 37 \times (4 \times 9)^x = 26 \times \left(9^2\right)^x$$
$$3 \times \left(4^x\right)^2 + 37 \times 4^x \times 9^x - 26 \times \left(9^x\right)^2 = 0.$$

Let $a = 4^x$ and $b = 9^x$. Then a and $b > 0$ for all $x \in \mathbb{R}$.
The given equation can be written as $3a^2 + 37ab - 26b^2 = 0$.

It follows that
$$(3a - 2b)(a + 13b) = 0.$$
Then $3a - 2b = 0$ since $a + 13b > 0$.
We obtain $3a = 2b$. Hence, $3 \times 4^x = 2 \times 9^x$. Then $\left(\dfrac{4}{9}\right)^x = \dfrac{2}{3}$.
It turns out that $\left(\dfrac{2}{3}\right)^{2x} = \dfrac{2}{3}$.
Then $2x = 1$. As a result, $x = \dfrac{1}{2}$.

Therefore, $x = \dfrac{1}{2}$.

6. $3^{2x^2+6x-9} + 4 \times 15^{x^2+3x-5} = 3 \times 5^{2x^2+6x-9}$
We have
$$3^{2x^2+6x-9} + 4 \times 15^{x^2+3x-5} = 3 \times 5^{2x^2+6x-9}$$
$$3 \times 3^{2x^2+6x-10} + 4 \times (3 \times 5)^{x^2+3x-5} = 3 \times 5 \times 5^{2x^2+6x-10}$$
$$3 \times 3^{2(x^2+3x-5)} + 4 \times 3^{x^2+3x-5} \times 5^{x^2+3x-5} - 15 \times 5^{2(x^2+3x-5)} = 0.$$

Let $a = 3^{x^2+3x-5}$ and $b = 5^{x^2+3x-5}$. Then a and $b > 0$ for all $x \in \mathbb{R}$. It follows that $3a^2 + 4ab - 15b^2 = 0$.
Then $(3a - 5b)(a + 3b) = 0$.
By knowing that $a + 3b > 0$, we obtain $3a - 5b = 0$.
Then $3a = 5b$.
As a result, $3 \times 3^{x^2+3x-5} = 5 \times 5^{x^2+3x-5}$.
It follows that $3^{x^2+3x-4} = 5^{x^2+3x-4}$.
Then $x^2 + 3x - 4 = 0$.
Since $a + b + c = 1 + 3 + (-4) = 0$, we obtain
$$x_1 = 1 \quad \text{and} \quad x_2 = \frac{c}{a} = \frac{-4}{1} = -4.$$

Therefore, $x \in \{-4, 1\}$.

7. $27^x + 12^x = 2 \times 8^x$
We have
$$27^x + 12^x = 2 \times 8^x$$

$$\left(3^3\right)^x + \left(2^2 \times 3\right)^x - 2 \times \left(2^3\right)^x = 0$$
$$\left(3^x\right)^3 + \left(2^x\right)^2 \times 3^x - 2 \times \left(2^x\right)^3 = 0.$$

Let $a = 3^x$ and $b = 2^x$. It follows that a and $b > 0$ for all $x \in \mathbb{R}$. The given equation can be written as

$$a^3 + ab^2 - 2b^3 = 0$$

or

$$\left(a^3 - b^3\right) + \left(ab^2 - b^3\right) = 0.$$

We obtain

$$(a - b)\left(a^2 + ab + b^2\right) + b^2(a - b) = 0$$

or

$$(a - b)\left(a^2 + ab + b^2 + b^2\right) = 0.$$

Then $(a - b)\left(a^2 + ab + 2b^2\right) = 0$.
By knowing that $a, b > 0$, it follows that $a^2 + ab + 2b^2 > 0$.
Then $a - b = 0$ or $a = b$.
As a result, $3^x = 2^x$. It turns out that $x = 0$.
$\boxed{\text{Therefore, } x = 0.}$

8. $5 \times 2^{3x-3} - 3 \times 2^{5-3x} + 7 = 0$
We have

$$5 \times 2^{3x-3} - 3 \times 2^{5-3x} + 7 = 0$$
$$5 \times 2^{3x-3} - 3 \times 2^2 \times 2^{3-3x} + 7 = 0$$
$$5 \times 2^{3x-3} - 12 \times 2^{-(3x-3)} + 7 = 0$$
$$5 \times 2^{3x-3} - \frac{12}{2^{3x-3}} + 7 = 0.$$

Let $t = 2^{3x-3}$. Then $t > 0$ for all $x \in \mathbb{R}$. The given equation can be written as

$$5t - \frac{12}{t} + 7 = 0$$

or

$$5t^2 + 7t - 12 = 0.$$

By knowing that $a + b + c = 5 + 7 - 12 = 0$, we obtain $t_1 = 1$ and $t_2 = \dfrac{c}{a} = -\dfrac{12}{5}$.

It follows that $t = 1$ since $t > 0$.
We obtain $2^{3x-3} = 1$. It implies that $3x - 3 = 0$.
Hence, $3x = 3$.
Therefore, $x = 1$.

9. $3^{2x^2-6x+3} + 6^{x^2-3x+1} = 2^{2x^2-6x+3}$

We have

$3^{2x^2-6x+3} + 6^{x^2-3x+1} = 2^{2x^2-6x+3}$

$3 \times 3^{2x^2-6x+2} + (2 \times 3)^{x^2-3x+1} - 2 \times 2^{2x^2-6x+2} = 0$

$3 \times 3^{2(x^2-3x+1)} + 2^{x^2-3x+1} \times 3^{x^2-3x+1} - 2 \times 2^{2(x^2-3x+1)} = 0$.

Divide both sides of the last equation by $2^{2(x^2-3x+1)}$, we obtain

$$3\left[\left(\frac{3}{2}\right)^{x^2-3x+1}\right]^2 + \left(\frac{3}{2}\right)^{x^2-3x+1} - 2 = 0.$$

Let $t = \left(\frac{3}{2}\right)^{x^2-3x+1}$. Then $t > 0$ for all $x \in \mathbb{R}$.
We obtain $3t^2 + t - 2 = 0$. Since $a + c = 3 - 2 = 1 = b$, it follows that

$$t_1 = -1 \text{ and } t_2 = -\frac{c}{a} = -\frac{-2}{3} = \frac{2}{3}.$$

By knowing that $t > 0$, we obtain $t = \frac{2}{3}$.
It turns out that $\left(\frac{2}{3}\right)^{x^2-3x+1} = \frac{2}{3}$.
We obtain $x^2 - 3x + 1 = 1$ or $x^2 - 3x = 0$.
Then $x(x - 3) = 0$.
It implies that $\begin{bmatrix} x = 0 \\ x - 3 = 0 \end{bmatrix}$ or $\begin{bmatrix} x = 0 \\ x = 3 \end{bmatrix}$.
Therefore, $x \in \{0, 3\}$.

Problem 18. Solve the following exponential equations:

1. $3^x + 4^x = 5^x$

2. $3^x + 4^x + 5^x = 6^x$

3. $10^x + 11^x + 12^x = 13^x + 14^x$

4. $6^x - 8^x - 9^x + 12^x = 0$

5. $2^x + 3^x - 4^x + 6^x - 9^x = 1$

Solution. Solve the following exponential equations:

1. $3^x + 4^x = 5^x$

   We have $3^x + 4^x = 5^x$. Divide both sides of the last equation by 5^x, we obtain

   $$\frac{3^x + 4^x}{5^x} = \frac{5^x}{5^x} \quad \text{or} \quad \frac{3^x}{5^x} + \frac{4^x}{5^x} = 1.$$

   It follows that $\left(\dfrac{3}{5}\right)^x + \left(\dfrac{4}{5}\right)^x = 1.$

   - For $x = 2$, we obtain

   $$\left(\frac{3}{5}\right)^x + \left(\frac{4}{5}\right)^x = \left(\frac{3}{5}\right)^2 + \left(\frac{4}{5}\right)^2$$
   $$= \frac{3^2 + 4^2}{5^2} = \frac{25}{25} = 1.$$

   It follows that $x = 2$.

   - For $x > 2$, we obtain

   $$\left(\frac{3}{5}\right)^x + \left(\frac{4}{5}\right)^x < \left(\frac{3}{5}\right)^2 + \left(\frac{4}{5}\right)^2 = 1.$$

   In this case, the equation has no solutions.

   - For $x < 2$, we obtain

   $$\left(\frac{3}{5}\right)^x + \left(\frac{4}{5}\right)^x > \left(\frac{3}{5}\right)^2 + \left(\frac{4}{5}\right)^2 = 1.$$

   In this case, the equation has no solutions.

   Therefore, $x = 2$.

2. $3^x + 4^x + 5^x = 6^x$
   Divide both sides of the given equation by 6^x, we obtain
   $$\frac{3^x + 4^x + 5^x}{6^x} = \frac{6^x}{6^x}$$
   or
   $$\left(\frac{3}{6}\right)^x + \left(\frac{4}{6}\right)^x + \left(\frac{5}{6}\right)^x = 1.$$
   - If $x = 3$, we obtain
   $$\left(\frac{3}{6}\right)^x + \left(\frac{4}{6}\right)^x + \left(\frac{5}{6}\right)^x = \left(\frac{3}{6}\right)^3 + \left(\frac{4}{6}\right)^3 + \left(\frac{5}{6}\right)^3$$
   $$= \frac{3^3}{6^3} + \frac{4^3}{6^3} + \frac{5^3}{6^3}$$
   $$= \frac{216}{216} = 1.$$
   Consequently, $x = 3$.
   - If $x > 3$, we obtain
   $$\left(\frac{3}{6}\right)^x + \left(\frac{4}{6}\right)^x + \left(\frac{5}{6}\right)^x < \left(\frac{3}{6}\right)^3 + \left(\frac{4}{6}\right)^3 + \left(\frac{5}{6}\right)^3 = 1.$$
   In this case, the equation has no solutions.
   - If $x < 3$, we obtain
   $$\left(\frac{3}{6}\right)^x + \left(\frac{4}{6}\right)^x + \left(\frac{5}{6}\right)^x > \left(\frac{3}{6}\right)^3 + \left(\frac{4}{6}\right)^3 + \left(\frac{5}{6}\right)^3 = 1.$$
   In this case, the equation has no solutions.

   $\boxed{\text{Therefore, } x = 3.}$

3. $10^x + 11^x + 12^x = 13^x + 14^x$
   If $x = 2$, we obtain
   $$10^x + 11^x + 12^x = 10^2 + 11^2 + 12^2$$
   $$= 100 + 121 + 144$$
   $$= 169 + 196$$
   $$= 13^2 + 14^2$$

$$= 13^x + 14^x.$$

Hence, $x = 2$ is a solution of the equation.
Divide both sides of the last equation by 13^x, we obtain

$$\frac{10^x}{13^x} + \frac{11^x}{13^x} + \frac{12^x}{13^x} = \frac{13^x}{13^x} + \frac{14^x}{13^x}$$

or

$$\left(\frac{10}{13}\right)^x + \left(\frac{11}{13}\right)^x + \left(\frac{12}{13}\right)^x = 1 + \left(\frac{14}{13}\right)^x. \quad (1)$$

We see that $f(x) = \left(\frac{10}{13}\right)^x + \left(\frac{11}{13}\right)^x + \left(\frac{12}{13}\right)^x$ is a decreasing function while $g(x) = 1 + \left(\frac{14}{13}\right)^x$ is an increasing function.
Hence, the equation (1) has a unique solution.
$\boxed{\text{Therefore, } x = 2 \text{ is the solution of the equation.}}$

4. $6^x - 8^x - 9^x + 12^x = 0$
   We have

$$6^x - 8^x - 9^x + 12^x = 0$$
$$(6^x - 9^x) - (8^x - 12^x) = 0$$
$$(2^x \times 3^x - 3^x \times 3^x) - (2^x \times 4^x - 3^x \times 4^x) = 0$$
$$3^x(2^x - 3^x) - 4^x(2^x - 3^x) = 0$$
$$(2^x - 3^x)(3^x - 4^x) = 0.$$

It follows that $\begin{bmatrix} 2^x - 3^x = 0 \\ 3^x - 4^x = 0 \end{bmatrix}$ or $\begin{bmatrix} 2^x = 3^x \\ 3^x = 4^x \end{bmatrix}$.

$\boxed{\text{Therefore, } x = 0 \text{ is the solution of the equation.}}$

5. $2^x + 3^x - 4^x + 6^x - 9^x = 1$
   We have

$$2^x + 3^x - 4^x + 6^x - 9^x = 1$$
$$2^x + 3^x - (2^2)^x + (2 \times 3)^x - (3^2)^x = 1$$
$$2^x + 3^x - (2^x)^2 + 2^x \times 3^x - (3^x)^2 = 1$$
$$(2^x)^2 + (3^x)^2 - 2^x \times 3^x - 2^x - 3^x + 1 = 0.$$

Let $a = 2^x$ and $b = 3^x$. It follows that a and $b > 0$ for all $x \in \mathbb{R}$.

The given equation can be written as $a^2+b^2-ab-a-b-1=0$. Multiply both sides of the last equation by 2, we obtain
$$2a^2 + 2b^2 - 2ab - 2a - 2b + 2 = 0.$$
Then
$$\left(a^2 - 2ab + b^2\right) + \left(a^2 - 2a + 1\right) + \left(b^2 - 2b + 1\right) = 0$$
or
$$(a-b)^2 + (a-1)^2 + (b-1)^2 = 0.$$

As a result, we obtain $\begin{cases} a - b = 0 \\ a - 1 = 0 \\ b - 1 = 0 \end{cases}$. Then $a = b = 1$.

Hence, $2^x = 3^x = 1$.

$\boxed{\text{Therefore, } x = 0.}$

Problem 19. Solve the following exponential equations:

1. $12^x - 3^x - 4^x + 1 = 0$
2. $6^x + 2^x - 3^x - 1 = 0$
3. $20^x - 4^x + 5^x - 1 = 0$
4. $30^x - 6 \times 5^x - 6^x + 6 = 0$

Solution. Solve the following exponential equations:

1. $12^x - 3^x - 4^x + 1 = 0$
   Let $a = 3^x$ and $b = 4^x$. Then $a, b > 0$ for all $x \in \mathbb{R}$. The given equation can be written as
   $$3^x \times 4^x - 3^x - 4^x + 1 = 0$$
   $$ab - a - b + 1 = 0$$
   $$a(b-1) - (b-1) = 0$$
   $$(a-1)(b-1) = 0.$$
   Then $\begin{bmatrix} a - 1 = 0 \\ b - 1 = 0 \end{bmatrix}$. It follows that $\begin{bmatrix} a = 1 \\ b = 1 \end{bmatrix}$.

   - If $a = 1$, we obtain $3^x = 1 = 3^0$. Then $x = 0$.

- If $b = 1$, we obtain $4^x = 1 = 4^0$. It follows that $x = 0$.

Therefore, $x = 0$.

2. $6^x + 2^x - 3^x - 1 = 0$
   Let $a = 2^x$ and $b = 3^x$. Then $a, b > 0$ for all $x \in \mathbb{R}$.
   The given equation can be written as

   $$6^x + 2^x - 3^x - 1 = 0$$
   $$2^x \times 3^x + 2^x - 3^x - 1 = 0$$
   $$ab + a - b - 1 = 0$$
   $$a(b + 1) - (b + 1) = 0$$
   $$(a - 1)(b + 1) = 0.$$

   Since $b + 1 > 0$, it follows that $a - 1 = 0$.
   Then $3^x = 1 = 3^0$. We obtain $x = 0$.

   Therefore, $x = 0$.

3. $20^x - 4^x + 5^x - 1 = 0$
   Let $a = 4^x$ and $b = 5^x$. Then $a, b > 0$ for all $x \in \mathbb{R}$. The given equation can be written as

   $$20^x - 4^x + 5^x - 1 = 0$$
   $$4^x \times 5^x - 4^x + 5^x - 1 = 0$$
   $$ab - a + b - 1 = 0$$
   $$a(b - 1) + (b - 1) = 0$$
   $$(a + 1)(b - 1) = 0.$$

   Since $a + 1 > 0$, it follows that $b - 1 = 0$. Then $b = 1$.
   We obtain $4^x = 1 = 4^0$. Hence, $x = 0$.

   Therefore, $x = 0$.

4. $30^x - 6 \times 5^x - 6^x + 6 = 0$
   Let $a = 5^x$ and $b = 6^x$. Then $a, b > 0$ for all $x \in \mathbb{R}$. The given equation can be written as

   $$30^x - 6 \times 5^x - 6^x + 6 = 0$$
   $$5^x \times 6^x - 6 \times 5^x - 6^x + 6 = 0$$
   $$ab - 6a - b + 6 = 0$$
   $$a(b - 6) - (b - 6) = 0$$

Chapter 5. Solutions

$$(a-1)(b-6) = 0.$$

Hence, $\begin{bmatrix} a-1=0 \\ b-6=0 \end{bmatrix}$. It follows that $\begin{bmatrix} a=1 \\ b=6 \end{bmatrix}$.

- For $a = 1$, we obtain $5^x = 1 = 5^0$. Then $x = 0$.
- For $b = 6$, we obtain $6^x = 6$. It follows that $x = 1$.

Therefore, $x \in \{0, 1\}$.

Problem 20. Solve the following exponential inequalities:

1. $2^x > 4$
2. $2^x < 4$
3. $0.1^x > 100$
4. $\left(\dfrac{1}{5}\right)^x > \sqrt[3]{0.04}$
5. $0.3^x > \dfrac{100}{9}$
6. $15^x > \dfrac{1}{\sqrt[4]{15^3}}$
7. $2^x < \dfrac{1}{64}$
8. $11^x > \sqrt[5]{11}$
9. $3^x < \dfrac{1}{9\sqrt{3}}$

Solution. Solve the following exponential inequalities:

1. $2^x > 4$
   We have $2^x > 4$. Then $2^x > 2^2$.
   Therefore, $x > 2$.

2. $2^x < 4$
   We have $2^x < 4$. Then $2^x < 2^2$.
   Therefore, $x < 2$.

3. $0.1^x > 100$
   We have $0.1^x > 100$. Then $\left(\dfrac{1}{10}\right)^x > 10^2$. It follows that $\left(\dfrac{1}{10}\right)^x > \left(\dfrac{1}{10}\right)^{-2}$.
   Therefore, $x < -2$.

4. $\left(\dfrac{1}{5}\right)^x > \sqrt[3]{0.04}$
   We have $\left(\dfrac{1}{5}\right)^x > \sqrt[3]{0.04}$. Then $\left(\dfrac{1}{5}\right)^x > 0.04^{\frac{1}{3}}$.

It follows that $\left(\dfrac{1}{5}\right)^x > \left(\dfrac{4}{100}\right)^{\frac{1}{3}}$.

Hence, $\left(\dfrac{1}{5}\right)^x > \left(\dfrac{1}{25}\right)^{\frac{1}{3}}$. We obtain $\left(\dfrac{1}{5}\right)^x > \left(\dfrac{1}{5}\right)^{\frac{2}{3}}$.

Therefore, $x < \dfrac{2}{3}$.

5. $0.3^x > \dfrac{100}{9}$

We have $0.3^x > \dfrac{100}{9}$. Then $\left(\dfrac{3}{10}\right)^x > \left(\dfrac{10}{3}\right)^2$.

It follows that $\left(\dfrac{3}{10}\right)^x > \left(\dfrac{3}{10}\right)^{-2}$.

Therefore, $x < -2$.

6. $15^x > \dfrac{1}{\sqrt[4]{15^3}}$

We have $15^x > \dfrac{1}{\sqrt[4]{15^3}}$. Then $15^x > \dfrac{1}{15^{\frac{3}{4}}}$. It follows that $15^x > 15^{-\frac{3}{4}}$.

Therefore, $x > -\dfrac{3}{4}$.

7. $2^x < \dfrac{1}{64}$

We have $2^x < \dfrac{1}{64}$. Then $2^x < \dfrac{1}{2^6}$. It follows that $2^x < 2^{-6}$.

Therefore, $x < -6$.

8. $11^x > \sqrt[5]{11}$

We have $11^x > \sqrt[5]{11}$. Then $11^x > 11^{\frac{1}{5}}$.

Therefore, $x > \dfrac{1}{5}$.

9. $3^x < \dfrac{1}{9\sqrt{3}}$

We have $3^x < \dfrac{1}{3^2 \times 3^{\frac{1}{2}}}$. Then $3^x < \dfrac{1}{3^{2+\frac{1}{2}}}$.

It follows that $3^x < \dfrac{1}{3^{\frac{5}{2}}}$. We obtain $3^x < 3^{-\frac{5}{2}}$.

Therefore, $x < -\dfrac{5}{2}$.

Problem 21. Solve the following exponential inequalities:

1. $25^{-4x} < 5\sqrt{5}$
2. $\left(\dfrac{2}{3}\right)^{-3x} < \dfrac{16}{81}$
3. $16^x > 0.125$
4. $\left(\dfrac{1}{49}\right)^{6x} > 7\sqrt[3]{49}$
5. $2^x > -5$
6. $2^x < -5$
7. $100^{x+1} < 10000$
8. $10^{x-2} \geq 0.01$
9. $27^x \times 3^{1-x} < \dfrac{1}{3}$

Solution. Solve the following exponential inequalities:

1. $25^{-4x} < 5\sqrt{5}$
   We have $25^{-4x} < 5\sqrt{5}$. Then $\left(5^2\right)^{-4x} < 5 \times 5^{\frac{1}{2}}$. It follows that $5^{-8x} < 5^{1+\frac{1}{2}}$.
   Then $5^{-8x} < 5^{\frac{3}{2}}$.
   We obtain $-8x < \dfrac{3}{2}$.
   Therefore, $x > -\dfrac{3}{16}$.

2. $\left(\dfrac{2}{3}\right)^{-3x} < \dfrac{16}{81}$
   We have $\left(\dfrac{2}{3}\right)^{-3x} < \dfrac{16}{81}$. Then $\left(\dfrac{2}{3}\right)^{-3x} < \left(\dfrac{2}{3}\right)^4$.
   We obtain $-3x > 4$.
   Therefore, $x < -\dfrac{4}{3}$.

3. $16^x > 0.125$
   We have $16^x > 0.125$. Then $16^x > \dfrac{125}{1000}$. It follows that $4^{2x} > \dfrac{1}{4}$.
   We obtain $4^{2x} > 4^{-1}$. Hence, $2x > -1$.
   Therefore, $x > -\dfrac{1}{2}$.

142

4. $\left(\dfrac{1}{49}\right)^{6x} > 7\sqrt[3]{49}$

   We have $\left(\dfrac{1}{49}\right)^{6x} > 7\sqrt[3]{49}$. Then $\left(\dfrac{1}{7^2}\right)^{6x} > 7(7^2)^{\frac{1}{3}}$.
   It follows that $\left(\dfrac{1}{7}\right)^{12x} > 7^{1+\frac{2}{3}}$. We obtain $\left(\dfrac{1}{7}\right)^{12x} > \left(\dfrac{1}{7}\right)^{-\frac{5}{3}}$.
   As a result, $12x < -\dfrac{5}{3}$.
   $\boxed{\text{Therefore, } x < -\dfrac{5}{36}.}$

5. $2^x > -5$
   Since $2^x > 0$ for all $x \in \mathbb{R}$, then $2^x > -5$ for all $x \in \mathbb{R}$.
   $\boxed{\text{Therefore, } x \in \mathbb{R}.}$

6. $2^x < -5$
   Since $2^x > 0$ for all $x \in \mathbb{R}$, there is no x that satisfies $2^x < -5$.
   $\boxed{\text{Therefore, the inequality has no solutions.}}$

7. $100^{x+1} < 10000$
   We have $100^{x+1} < 10000$. Then $100^{x+1} < 100^2$.
   It follows that $x + 1 < 2$. Then $x < 1$.
   $\boxed{\text{Therefore, } x < 1.}$

8. $10^{x-2} \geq 0.01$
   We have $10^{x-2} \geq 0.01$. Then $10^{x-2} \geq 10^{-2}$.
   It follows that $x - 2 \geq -2$. We obtain $x \geq 0$.
   $\boxed{\text{Therefore, } x \geq 0.}$

9. $27^x \times 3^{1-x} < \dfrac{1}{3}$
   We have $27^x \times 3^{1-x} < \dfrac{1}{3}$. Then $3^{3x} \times 3^{1-x} < 3^{-1}$.
   It follows that $3^{3x+1-x} < 3^{-1}$. Then $3^{2x+1} < 3^{-1}$.
   Hence, $2x + 1 < -1$. We obtain $2x < -2$.
   $\boxed{\text{Therefore, } x < -1.}$

Problem 22. Solve the following exponential inequalities:

Chapter 5. Solutions

1. $0.1^x > 10$
2. $12^x < \sqrt[3]{144}$
3. $0.4^x < \dfrac{125}{8}$
4. $2^x > \sqrt{2^3}$
5. $10^x > \dfrac{1}{\sqrt{10}}$
6. $0.2^x < 25$
7. $5^x < \sqrt[3]{25}$
8. $0.7^x > \dfrac{100}{49}$
9. $\left(\dfrac{1}{3}\right)^x > \sqrt[3]{\dfrac{1}{9}}$

Solution. Solve the following exponential inequalities:

1. $0.1^x > 10$
   We have $0.1^x > 10$. Then $\left(\dfrac{1}{10}\right)^x > 10$.
   It follows that $\left(\dfrac{1}{10}\right)^x > \left(\dfrac{1}{10}\right)^{-1}$.
   $\boxed{\text{Therefore, } x < -1.}$

2. $12^x < \sqrt[3]{144}$
   We have $12^x < \sqrt[3]{144}$. Then $12^x < 144^{\frac{1}{3}}$.
   It follows that $12^x < 12^{\frac{2}{3}}$.
   $\boxed{\text{Therefore, } x < \dfrac{2}{3}.}$

3. $0.4^x < \dfrac{125}{8}$
   We have $0.4^x < \dfrac{125}{8}$. Then $\left(\dfrac{4}{10}\right)^x < \left(\dfrac{5}{2}\right)^3$.
   It follows that $\left(\dfrac{2}{5}\right)^x < \left(\dfrac{2}{5}\right)^{-3}$.
   $\boxed{\text{Therefore, } x > -3.}$

4. $2^x > \sqrt{2^3}$
   We have $2^x > \sqrt{2^3}$. Then $2^x > 2^{\frac{3}{2}}$.
   $\boxed{\text{Therefore, } x > \dfrac{3}{2}.}$

5. $10^x > \dfrac{1}{\sqrt{10}}$
   We have $10^x > \dfrac{1}{\sqrt{10}}$. Then $10^x > \dfrac{1}{10^{\frac{1}{2}}}$.

It follows that $10^x > 10^{-\frac{1}{2}}$.

$\boxed{\text{Therefore, } x > -\frac{1}{2}.}$

6. $0.2^x < 25$

We have $0.2^x < 25$. Then $\left(\frac{2}{10}\right)^x < 25$.

It follows that $\left(\frac{1}{5}\right)^x < 5^2$.

We obtain $\left(\frac{1}{5}\right)^x < \left(\frac{1}{5}\right)^{-2}$.

$\boxed{\text{Therefore, } x > -2.}$

7. $5^x < \sqrt[3]{25}$

We have $5^x < \sqrt[3]{25}$. Then $5^x < 5^{\frac{2}{3}}$.

$\boxed{\text{Therefore, } x < \frac{2}{3}.}$

8. $0.7^x > \frac{100}{49}$

We have $0.7^x > \frac{100}{49}$. Then $\left(\frac{7}{10}\right)^x > \left(\frac{10}{7}\right)^2$.

It follows that $\left(\frac{7}{10}\right)^x > \left(\frac{7}{10}\right)^{-2}$.

$\boxed{\text{Therefore, } x < -2.}$

9. $\left(\frac{1}{3}\right)^x > \sqrt[3]{\frac{1}{9}}$

We have $\left(\frac{1}{3}\right)^x > \sqrt[3]{\frac{1}{9}}$. Then $\left(\frac{1}{3}\right)^x > \sqrt[3]{\left(\frac{1}{3}\right)^2}$.

It follows that $\left(\frac{1}{3}\right)^x > \left(\frac{1}{3}\right)^{\frac{2}{3}}$.

$\boxed{\text{Therefore, } x < \frac{2}{3}.}$

Problem 23. Solve the following exponential inequalities:

1. $25^x > 125^{3x-2}$
2. $4^{-x+\frac{1}{2}} - 7 \times 2^{-x} - 4 < 0$

Chapter 5. Solutions

3. $5^{2x-1} \times 7^{3x+2} \leq 7^{2x-1} \times 5^{3x+2}$

4. $2^{x+2} - 2^{x+3} - 2^{x+4} > 5^{x+1} - 5^{x+2}$

5. $0.1^{4x^2-2x-2} \leq 0.1^{2x-3}$

6. $2^{x^2} \times 5^{x^2} < 10^{-3}(10^{3-x})^2$

7. $2^{9x-x^3} > 1$

8. $2^{9x-x^3} < 1$

9. $2^{2x^2-6x+3} + 6^{x^2-3x+1} \geq 3^{2x^2-6x+3}$

Solution. Solve the following exponential inequalities:

1. $25^x > 125^{3x-2}$
   We have $25^x > 125^{3x-2}$. Then $5^{2x} > 5^{3(3x-2)}$.
   It follows that $5^{2x} > 5^{9x-6}$.
   Hence, $2x > 9x - 6$. As a result, $7x < 6$.
   $\boxed{\text{Therefore, } x < \dfrac{6}{7}.}$

2. $4^{-x+\frac{1}{2}} - 7 \times 2^{-x} - 4 < 0$
   We have $4^{-x+\frac{1}{2}} - 7 \times 2^{-x} - 4 < 0$. Then
   $$4^{-x} \times 4^{\frac{1}{2}} - 7 \times 2^{-x} - 4 < 0$$
   or
   $$2 \times 2^{-2x} - 7 \times 2^{-x} - 4 < 0.$$

   Let $t = 2^{-x}$. Then $t > 0$ for all $x \in \mathbb{R}$. The given equation can be written as
   $$2t^2 - 7t - 4 < 0.$$

   It follows that $(2t+1)(t-4) < 0$. Since $2t+1 > 0$, we obtain
   $$t - 4 < 0 \text{ or } t < 4.$$

   Hence, $2^{-x} < 4$. It implies that $2^{-x} < 2^2$.
   Then $-x < 2$.
   $\boxed{\text{Therefore, } x > -2.}$

146

3. $5^{2x-1} \times 7^{3x+2} \leq 7^{2x-1} \times 5^{3x+2}$
   We have $5^{2x-1} \times 7^{3x+2} \leq 7^{2x-1} \times 5^{3x+2}$.
   Then $\left(\dfrac{5}{7}\right)^{2x-1} \leq \left(\dfrac{5}{7}\right)^{3x+2}$.
   It follows that $2x - 1 \geq 3x + 2$. We obtain $3x - 2x \leq -2 - 1$.
   Therefore, $x \leq -3$.

4. $2^{x+2} - 2^{x+3} - 2^{x+4} > 5^{x+1} - 5^{x+2}$
   We have $2^{x+2} - 2^{x+3} - 2^{x+4} > 5^{x+1} - 5^{x+2}$.
   Then
   $$2^2 \times 2^x - 2^3 \times 2^x - 2^4 \times 2^x > 5 \times 5^x - 5^2 \times 5^x$$
   or
   $$4 \times 2^x - 8 \times 2^x - 16 \times 2^x > 5 \times 5^x - 25 \times 5^x.$$
   It follows that $-20 \times 2^x > -20 \times 5^x$. We obtain $2^x < 5^x$.
   It implies that $\left(\dfrac{2}{5}\right)^x < 1$. We obtain $\left(\dfrac{2}{5}\right)^x < \left(\dfrac{2}{5}\right)^0$.
   Therefore, $x > 0$.

5. $0.1^{4x^2 - 2x - 2} \leq 0.1^{2x-3}$
   We have $0.1^{4x^2 - 2x - 2} \leq 0.1^{2x-3}$. Then
   $$4x^2 - 2x - 2 \geq 2x - 3$$
   or
   $$4x^2 - 4x + 1 \geq 0$$
   It follows that $(2x - 1)^2 \geq 0$. It turns out that $x \in \mathbb{R}$.
   Therefore, $x \in \mathbb{R}$.

6. $2^{x^2} \times 5^{x^2} < 10^{-3}\left(10^{3-x}\right)^2$
   We have $2^{x^2} \times 5^{x^2} < 10^{-3}\left(10^{3-x}\right)^2$. Then
   $$10^{x^2} < 10^{-3} \times 10^{2(3-x)}$$
   or
   $$10^{x^2} < 10^{-3+6-2x}.$$
   It follows that $10^{x^2} < 10^{3-2x}$.

Hence, $x^2 < 3 - 2x$. We obtain $x^2 + 2x - 3 < 0$.
If $x^2 + 2x - 3 = 0$, it follows that $(x-1)(x+3) = 0$.
Then $\left[\begin{array}{l} x - 1 = 0 \\ x + 3 = 0 \end{array} \right.$. It implies that $\left[\begin{array}{l} x = 1 \\ x = -3 \end{array} \right.$.

x	$-\infty$		-3		1		$+\infty$
$x^2 + 2x - 3$		$+$	0	$-$	0	$+$	

Therefore, $x \in (-3, 1)$.

7. $2^{9x-x^3} > 1$
We have $2^{9x-x^3} > 1$. Then $2^{9x-x^3} > 2^0$.
It follows that $9x - x^3 > 0$. We obtain $x^3 - 9x < 0$.
Then $x\left(x^2 - 9\right) < 0$.
If $x\left(x^2 - 9\right) = 0$, we obtain $x(x-3)(x+3) = 0$.
It implies that $\left[\begin{array}{l} x = 0 \\ x - 3 = 0 \\ x + 3 = 0 \end{array} \right.$ or $\left[\begin{array}{l} x = 0 \\ x = 3 \\ x = -3 \end{array} \right.$.

x	$-\infty$		-3		0		3		$+\infty$
x		$-$		$-$	0	$+$		$+$	
$x - 3$		$-$		$-$		$-$	0	$+$	
$x + 3$		$-$	0	$+$		$+$		$+$	
$x(x-3)(x+3)$		$-$	0	$+$	0	$-$	0	$+$	

Therefore, $x \in (-\infty, -3) \cup (0, 3)$.

8. $2^{2x^2-6x+3} + 6^{x^2-3x+1} \geq 3^{2x^2-6x+3}$
We have $2^{2x^2-6x+3} + 6^{x^2-3x+1} \geq 3^{2x^2-6x+3}$.
Then

$$2 \times 2^{2x^2-6x+2} + 2^{x^2-3x+1} \times 3^{x^2-3x+1} - 3 \times 3^{2x^2-6x+2} \geq 0$$

or
$$2 \times 2^{2(x^2-3x+1)} + 2^{x^2-3x+1} \times 3^{x^2-3x+1} - 3 \times 3^{2(x^2-3x+1)} \geq 0.$$

Divide both sides of the inequality by $3^{2(x^2-3x+1)}$, we obtain

$$2\left[\left(\frac{2}{3}\right)^{x^2-3x+1}\right]^2 + \left(\frac{2}{3}\right)^{x^2-3x+1} - 3 \geq 0.$$

Let $t = \left(\frac{2}{3}\right)^{x^2-3x+1}$. We obtain $t > 0$ for all $x \in \mathbb{R}$. The given inequality can be written as $2t^2 + t - 3 \geq 0$.
It follows that $(t-1)(2t+3) \geq 0$.
Since $2t + 3 > 0$, it implies that $t - 1 \geq 0$. Then $t \geq 1$.
We obtain $\left(\frac{2}{3}\right)^{x^2-3x+1} \geq 1$.

It follows that $\left(\frac{2}{3}\right)^{x^2-3x+1} \geq \left(\frac{2}{3}\right)^0$.
Then $x^2 - 3x + 1 \geq 0$.
If $x^2 - 3x + 1 = 0$, we obtain the discriminant

$$\Delta = b^2 - 4ac$$
$$= (-3)^2 - 4(1)(1)$$
$$= 9 - 4 = 5.$$

We obtain
$$x_1 = \frac{-b + \sqrt{\Delta}}{2a} = \frac{-(-3) + \sqrt{5}}{2} = \frac{3 + \sqrt{5}}{2}$$

and
$$x_2 = \frac{-b - \sqrt{\Delta}}{2a} = \frac{-(-3) - \sqrt{5}}{2} = \frac{3 - \sqrt{5}}{2}.$$

x	$-\infty$		$\frac{3-\sqrt{5}}{2}$		$\frac{3+\sqrt{5}}{2}$		$+\infty$
$x^2 - 3x + 1$		$+$	0	$-$	0	$+$	

Therefore, $x \in \left(-\infty, \dfrac{3-\sqrt{5}}{2}\right] \cup \left[\dfrac{3+\sqrt{5}}{2}, +\infty\right).$

Problem 24. By definition, write the following expressions in logarithmic forms:

1. $4^3 = 64$
2. $3^5 = 243$
3. $7^3 = 343$
4. $10^{-2} = \dfrac{1}{100}$
5. $4^{-3} = \dfrac{1}{64}$
6. $x^y = k$

Solution. By definition, write the following expressions in logarithmic forms:

1. $4^3 = 64$ is equivalent to $3 = \log_4 64$.
2. $3^5 = 243$ is equivalent to $5 = \log_3 243$.
3. $7^3 = 343$ is equivalent to $3 = \log_7 343$.
4. $10^{-2} = \dfrac{1}{100}$ is equivalent to $-2 = \log \dfrac{1}{100}$.
5. $4^{-3} = \dfrac{1}{64}$ is equivalent to $-3 = \log_4 \dfrac{1}{64}$.
6. $x^y = k$ is equivalent to $y = \log_x k$.

Problem 25. Compute the following expressions:

1. $2^{\log_2 5}$
2. $2^{2\log_2 3}$
3. $\log_3 27$
4. $\log_{\sqrt{2}} 32$
5. $\log_{\sqrt{2}} \sqrt{8}$
6. $\log 0.01$
7. $\log_{2022} 1$
8. $\log_{\sqrt[4]{3}} \sqrt[4]{27}$
9. $\log 2 + \log 5$
10. $\log_2 10 - \log_2 5$

Solution. Compute the following expressions:

1. $2^{\log_2 5} = 5$
2. $2^{2\log_2 3} = 2^{\log_2 3^2} = 3^2 = 9$

3. $\log_3 27 = \log_3 3^3 = 3$

4. $\log_{\sqrt{2}} 32 = \log_{2^{\frac{1}{2}}} 2^5 = \frac{5}{\frac{1}{2}} \log_2 2 = 10$

5. $\log_{\sqrt{2}} \sqrt{8} = \log_{2^{\frac{1}{2}}} 2^{\frac{3}{2}} = \frac{\frac{3}{2}}{\frac{1}{2}} \log_2 2 = \frac{3}{2} \times 2 = 3$

6. $\log 0.01 = \log \frac{1}{100} = \log 10^{-2} = -2$

7. $\log_{2022} 1 = 0$

8. $\log_{\sqrt[4]{3}} \sqrt[4]{27} = \log_{3^{\frac{1}{4}}} 27^{\frac{1}{4}} = \frac{\frac{1}{4}}{\frac{1}{4}} \log_3 27 = \log_3 3^3 = 3$

9. $\log 2 + \log 5 = \log 10 = 1$

10. $\log_2 10 - \log_2 5 = \log_2 2 = 1$

Problem 26. Given three positive real numbers $a, b, x \neq 1$ such that $\log_a x = y$ and $\log_b x = z$. Find $\log_{ab} x$ in terms of y and z.

Solution. Find $\log_{ab} x$ in terms of y and z.
We have
$$\log_{ab} x = \frac{1}{\log_x ab}$$
$$= \frac{1}{\log_x a + \log_x b}$$
$$= \frac{1}{\frac{1}{\log_a x} + \frac{1}{\log_b x}}$$
$$= \frac{1}{\frac{1}{x} + \frac{1}{y}}$$
$$= \frac{1}{\frac{x+y}{xy}}$$
$$= \frac{xy}{x+y}$$

Therefore, $\log_{ab} x = \dfrac{xy}{x+y}$.

Problem 27. Prove that $\log_2\log_3\sqrt{3\sqrt{3\sqrt{3\sqrt{3}}}} = \log_2 15 - 4$.

Solution. Prove that $\log_2\log_3\sqrt{3\sqrt{3\sqrt{3\sqrt{3}}}} = \log_2 15 - 4$.
We have

$$\log_2\log_3\sqrt{3\sqrt{3\sqrt{3\sqrt{3}}}} = \log_2\log_3 3^{\frac{1}{2}+\frac{1}{4}+\frac{1}{8}+\frac{1}{16}}$$
$$= \log_2\log_3 3^{\frac{15}{16}}$$
$$= \log_2 \frac{15}{16}$$
$$= \log_2 15 - \log_2 16$$
$$= \log_2 15 - \log_2 2^4$$
$$= \log_2 15 - 4.$$

Therefore, $\log_2\log_3\sqrt{3\sqrt{3\sqrt{3\sqrt{3}}}} = \log_2 15 - 4$.

Problem 28. Compute $3^{2\log_3 2} + 4^{3\log_2 5} + 5^{\frac{1}{\log_2 5}}$.

Solution. Compute $3^{2\log_3 2} + 4^{3\log_2 5} + 5^{\frac{1}{\log_2 5}}$.
We have

$$3^{2\log_3 2} + 4^{3\log_2 5} + 5^{\frac{1}{\log_2 5}} = 3^{\log_3 2^2} + 4^{\log_2 5^3} + 5^{\log_5 2}$$
$$= 2^2 + \left(2^2\right)^{\log_2 5^3} + 2$$
$$= 4 + \left(2^{\log_2 125}\right)^2 + 2$$
$$= 6 + 125^2$$
$$= 15631.$$

Therefore, $3^{2\log_3 2} + 4^{3\log_2 5} + 5^{\frac{1}{\log_2 5}} = 15631$.

Problem 29. Given a and b are two positive numbers greater than 1 such that $\log_a b = 2$. Compute

1. $\log_{ab}\left(\dfrac{\sqrt[4]{a}}{\sqrt[3]{b}}\right)$
2. $\log_{\sqrt[3]{a}\sqrt[4]{b}}\left(\dfrac{a}{b}\right)$

Solution. Compute

1. $\log_{ab}\left(\dfrac{\sqrt[4]{a}}{\sqrt[3]{b}}\right)$

We have

$$\log_{ab}\left(\dfrac{\sqrt[4]{a}}{\sqrt[3]{b}}\right) = \log_{ab}\sqrt[4]{a} - \log_{ab}\sqrt[3]{b}$$

$$= \log_{ab}a^{\frac{1}{4}} - \log_{ab}b^{\frac{1}{3}}$$

$$= \frac{1}{4}\log_{ab}a - \frac{1}{3}\log_{ab}b$$

$$= \frac{1}{4\log_a ab} - \frac{1}{3\log_b ab}$$

$$= \frac{1}{4(\log_a a + \log_a b)} - \frac{1}{3(\log_b a + \log_b b)}$$

$$= \frac{1}{4(1 + \log_a b)} - \frac{1}{3(\log_b a + 1)}$$

$$= \frac{1}{4(1 + \log_a b)} - \frac{1}{3\left(1 + \dfrac{1}{\log_a b}\right)}.$$

Since $\log_a b = 2$, it follows that

$$\log_{ab}\left(\dfrac{\sqrt[4]{a}}{\sqrt[3]{b}}\right) = \frac{1}{4(1+2)} - \frac{1}{3\left(1+\frac{1}{2}\right)}$$

$$= \frac{1}{12} - \frac{2}{9}$$

$$= \frac{3-8}{36}$$

$$= -\frac{5}{36}.$$

Therefore, $\log_{ab}\left(\dfrac{\sqrt[4]{a}}{\sqrt[3]{b}}\right) = -\dfrac{5}{36}$.

2. $\log_{\sqrt[3]{a}\sqrt[4]{b}}\left(\dfrac{a}{b}\right)$

We have

$$\log_{\sqrt[3]{a}\sqrt[4]{b}}\left(\dfrac{a}{b}\right) = \log_{a^{\frac{1}{3}}b^{\frac{1}{4}}}\left(\dfrac{a}{b}\right)$$

$$= \log_{a^{\frac{1}{3}}b^{\frac{1}{4}}} a - \log_{a^{\frac{1}{3}}b^{\frac{1}{4}}} b$$

$$= \frac{1}{\log_a a^{\frac{1}{3}}b^{\frac{1}{4}}} - \frac{1}{\log_b a^{\frac{1}{3}}b^{\frac{1}{4}}}$$

$$= \frac{1}{\log_a a^{\frac{1}{3}} + \log_a b^{\frac{1}{4}}} - \frac{1}{\log_b a^{\frac{1}{3}} + \log_b b^{\frac{1}{4}}}$$

$$= \frac{1}{\frac{1}{3} + \frac{1}{4}\log_a b} - \frac{1}{\frac{1}{3}\log_b a + \frac{1}{4}}$$

$$= \frac{1}{\frac{1}{3} + \frac{1}{4}\log_a b} - \frac{1}{\frac{1}{3\log_a b} + \frac{1}{4}}.$$

Since $\log_a b = 2$, it follows that

$$\log_{\sqrt[3]{a}\sqrt[4]{b}}\left(\frac{a}{b}\right) = \frac{1}{\frac{1}{3} + \frac{1}{4} \times 2} - \frac{1}{\frac{1}{3 \times 2} + \frac{1}{4}}$$

$$= \frac{1}{\frac{1}{3} + \frac{1}{2}} - \frac{1}{\frac{1}{6} + \frac{1}{4}}$$

$$= \frac{1}{\frac{5}{6}} - \frac{1}{\frac{5}{12}}$$

$$= \frac{6}{5} - \frac{12}{5}$$

$$= -\frac{6}{5}.$$

$\boxed{\text{Therefore, } \log_{\sqrt[3]{a}\sqrt[4]{b}}\left(\frac{a}{b}\right) = -\frac{6}{5}.}$

Problem 30. Let a and b be two positive real numbers such that $a^2 + b^2 = 3ab$. Prove that $\log_5(a+b) = \frac{1}{2}(1 + \log_5 a + \log_5 b)$.

Solution. We have $a^2 + b^2 = 3ab$. Adding $2ab$ to both sides of the equation, we obtain

$$(a+b)^2 = 5ab.$$

It follows that

$$\log_5(a+b)^2 = \log_5 5ab$$

$$2\log_5(a+b) = \log_5 5 + \log_5 a + \log_5 b$$
$$2\log_5(a+b) = 1 + \log_5 a + \log_5 b.$$

Therefore, $\log_5(a+b) = \dfrac{1}{2}(1 + \log_5 a + \log_5 b).$

Problem 31. For all $x > 1$, prove that

$$\frac{1}{\log_2 x} + \frac{1}{\log_3 x} + \ldots + \frac{1}{\log_{2022} x} = \log_x 2022!.$$

Solution. Prove that $\dfrac{1}{\log_2 x} + \dfrac{1}{\log_3 x} + \ldots + \dfrac{1}{\log_{2022} x} = \log_x 2022!.$
Since $\log_a b = \dfrac{1}{\log_b a}$, it follows that

$$\frac{1}{\log_2 x} + \frac{1}{\log_3 x} + \ldots + \frac{1}{\log_{2022} x}$$
$$= \log_x 2 + \log_x 3 + \ldots + \log_x 2022$$
$$= \log_x (2 \times 3 \times \ldots \times 2022)$$
$$= \log_x 2022!.$$

Therefore, $\dfrac{1}{\log_2 x} + \dfrac{1}{\log_3 x} + \ldots + \dfrac{1}{\log_{2022} x} = \log_x 2022!.$

Problem 32. Given a, b and $c > 1$ such that $\log_a b + \log_b c + \log_c a = 1$. Compute
$$\frac{\log_a c}{\log_{ab} c} + \frac{\log_b a}{\log_{bc} a} + \frac{\log_c b}{\log_{ca} b}.$$

Solution. Compute $\dfrac{\log_a c}{\log_{ab} c} + \dfrac{\log_b a}{\log_{bc} a} + \dfrac{\log_c b}{\log_{ca} b}.$
We have

$$\frac{\log_a c}{\log_{ab} c} = \frac{\log_c ab}{\log_c a}$$
$$= \frac{\log_c a + \log_c b}{\log_c a}$$
$$= \frac{\log_c a}{\log_c a} + \frac{\log_c b}{\log_c a}$$
$$= 1 + \log_a b.$$

Similarly, $\dfrac{\log_b a}{\log_{bc} a} = 1 + \log_b c$ and $\dfrac{\log_c a}{\log_{ca} a} = 1 + \log_c a$.
We obtain

$$\dfrac{\log_a c}{\log_{ab} c} + \dfrac{\log_b a}{\log_{bc} a} + \dfrac{\log_c a}{\log_{ca} a}$$
$$= 1 + \log_a b + 1 + \log_b c + 1 + \log_c a$$
$$= 3 + (\log_a b + \log_b c + \log_c a)$$
$$= 3 + 1$$
$$= 4.$$

Therefore, $\dfrac{\log_a c}{\log_{ab} c} + \dfrac{\log_b a}{\log_{bc} a} + \dfrac{\log_c b}{\log_{ca} b} = 4.$

Problem 33. Given $\log_2 3 = x$ and $\log_3 4 = y$.
Compute $E = \dfrac{1 + xy}{1 - xy}$.

Solution. Compute E.
Since $\log_2 3 = x$ and $\log_3 4 = y$, it follows that

$$xy = \log_2 3 \times \log_3 4$$
$$= \log_2 3 \times \log_3 2^2$$
$$= 2\log_2 3 \times \dfrac{1}{\log_2 3}$$
$$= 2.$$

Then $E = \dfrac{1 + 2}{1 - 2} = -3$.

Therefore, $E = -3.$

Problem 34. Given a, b and $c > 1$. Prove that $\log_a b \times \log_b c = \log_a c$. Using induction, show that $\log_{x_1} x_2 \times \log_{x_2} x_3 \times ... \times \log_{x_{n-1}} x_n = \log_{x_1} x_n$ for all $n \geq 2$.

Solution. Prove that $\log_a b \times \log_b c = \log_a c$.
We have $\log_b c = \dfrac{\log_a c}{\log_a b}$.

Therefore, $\log_a b \times \log_b c = \log_a c.$

+ Show that $\log_{x_1} x_2 \times \log_{x_2} x_3 \times ... \times \log_{x_{n-1}} x_n = \log_{x_1} x_n$.
For $n = 2$, we obtain $\log_{x_1} x_2 \times \log_{x_2} x_3 = \log_{x_1} x_3$, which is true.

Suppose that $\log_{x_1} x_2 \times \log_{x_2} x_3 \times \ldots \times \log_{x_{n-1}} x_n = \log_{x_1} x_n$. We shall show that

$$\log_{x_1} x_2 \times \log_{x_2} x_3 \times \ldots \times \log_{x_n} x_{n+1} = \log_{x_1} x_{n+1}.$$

We have

$$\log_{x_1} x_2 \times \log_{x_2} x_3 \times \ldots \times \log_{x_n} x_{n+1} = \log_{x_1} x_n \times \log_{x_n} x_{n+1}$$
$$= \log_{x_1} x_{n+1}.$$

Therefore, $\log_{x_1} x_2 \times \log_{x_2} x_3 \times \ldots \times \log_{x_{n-1}} x_n = \log_{x_1} x_n.$

Problem 35. Given four real numbers a, b, c and $x > 1$. Prove that

$$\log_a x \log_b x + \log_b x \log_c x + \log_c x \log_a x = \frac{\log_a x \log_b x \log_c x}{\log_{abc} x}.$$

Solution. Prove that

$$\log_a x \log_b x + \log_b x \log_c x + \log_c x \log_a x = \frac{\log_a x \log_b x \log_c x}{\log_{abc} x}.$$

We have

$$\log_a x \log_b x + \log_b x \log_c x + \log_c x \log_a x$$
$$= \frac{1}{\log_x a \log_x b} + \frac{1}{\log_x b \log_x c} + \frac{1}{\log_x c \log_x a}$$
$$= \frac{\log_x a + \log_x b + \log_x c}{\log_x a \log_x b \log_x c}$$
$$= \frac{\log_x abc}{\log_x a \log_x b \log_x c}$$
$$= \frac{\log_a x \log_b x \log_c x}{\log_{abc} x}.$$

Therefore, $\log_a x \log_b x + \log_b x \log_c x + \log_c x \log_a x = \dfrac{\log_a x \log_b x \log_c x}{\log_{abc} x}.$

Problem 36. Suppose that x, y and z are three distinct positive real numbers greater than 1. If $\log_y x \log_z x + \log_x y \log_z y + \log_y z \log_x z = 3$, find xyz.

Solution. Find xyz.
We have

$$\log_y x \log_z x + \log_x y \log_z y + \log_y z \log_x z = 3$$

$$\frac{\log x \log x}{\log y \log z} + \frac{\log y \log y}{\log x \log z} + \frac{\log z \log z}{\log y \log x} = 3$$

$$\frac{\log^2 x}{\log y \log z} + \frac{\log^2 y}{\log x \log z} + \frac{\log^2 z}{\log y \log x} = 3$$

$$\log^3 x + \log^3 y + \log^3 z = 3 \log x \log y \log z.$$

Using the fact that if $a^3 + b^3 + c^3 = 3abc$, we obtain $a+b+c = 0$, it follows that $\log x + \log y + \log z = 0$. Then $\log xyz = 0$. Therefore, $xyz = 1$.

Remark 3. In the above computation, we denote $\log x = \log_{10} x$.

Problem 37. Given a and b are two positive real numbers such that $\log_{16} a + \log_8 b = \dfrac{1}{4}$ and $\log_{16} b + \log_8 a = \dfrac{1}{5}$. Compute ab.

Solution. Compute ab.
We have

$$\log_{16} a + \log_8 b = \frac{1}{4}$$

$$\log_{2^4} a + \log_{2^3} b = \frac{1}{4}$$

$$\frac{1}{4}\log_2 a + \frac{1}{3}\log_2 b = \frac{1}{4}. \qquad (1)$$

and

$$\log_{16} b + \log_8 a = \frac{1}{5}$$

$$\log_{2^4} b + \log_{2^3} a = \frac{1}{5}$$

$$\frac{1}{4}\log_2 b + \frac{1}{3}\log_2 a = \frac{1}{5}. \qquad (2)$$

Adding (1) and (2), we obtain

$$\left(\frac{1}{4} + \frac{1}{3}\right)\log_2 a + \left(\frac{1}{4} + \frac{1}{3}\right)\log_2 b = \frac{1}{4} + \frac{1}{5}$$

$$\frac{7}{12}\log_2 a + \frac{7}{12}\log_2 b = \frac{9}{20}$$
$$\frac{7}{12}(\log_2 a + \log_2 b) = \frac{9}{20}$$
$$\log_2 ab = \frac{9}{20} \times \frac{12}{7}$$
$$\log_2 ab = \frac{27}{35}.$$

Therefore, $ab = 2^{\frac{27}{35}}$.

Problem 38. Given x, y and z are three positive real number greater than 1 such that $a^x = b^y = c^z = k$. Prove that

$$x\log_x a + y\log_y b + z\log_z c = \frac{\log_k x \log_k y + \log_k y \log_k z + \log_k z \log_k x}{\log_k x \log_k y \log_k z}.$$

Solution. Prove that

$$x\log_x a + y\log_y b + z\log_z c = \frac{\log_k x \log_k y + \log_k y \log_k z + \log_k z \log_k x}{\log_k x \log_k y \log_k z}.$$

We have $a^x = b^y = c^z = k$. By definition, we obtain $x = \log_a k$, $y = \log_b k$ and $z = \log_c k$.
Hence,

$$x\log_x a + y\log_y b + z\log_z c$$
$$= \log_x a \log_a k + \log_y b \log_b k + \log_z c \log_c k$$
$$= \log_x k + \log_y k + \log_z k$$
$$= \frac{1}{\log_k x} + \frac{1}{\log_k y} + \frac{1}{\log_k z}$$
$$= \frac{\log_k x \log_k y + \log_k y \log_k z + \log_k z \log_k x}{\log_k x \log_k y \log_k z}.$$

Therefore, $x\log_x a + y\log_y b + z\log_z c$
$$= \frac{\log_k x \log_k y + \log_k y \log_k z + \log_k z \log_k x}{\log_k x \log_k y \log_k z}.$$

Problem 39. Given three positive real numbers a, b and $c > 1$. Prove that $a^{\log_b c} = c^{\log_b a}$.

Solution. Prove that $a^{\log_b c} = c^{\log_b a}$.
Let $x = a^{\log_b c}$. Then $\log_a x = \log_b c$. It follows that $\dfrac{\log x}{\log a} = \dfrac{\log c}{\log b}$.
Hence, $\dfrac{\log x}{\log c} = \dfrac{\log a}{\log b}$.
We obtain $\log_c x = \log_b a$. Then $x = c^{\log_b}$.
$\boxed{\text{Therefore, } a^{\log_b c} = c^{\log_b a}.}$

Problem 40. Solve the following logarithmic equations:

1. $\log_2(3x) = 1$
2. $\log_2 x + \log_2 4 = \log_2 5$
3. $3^{2\log_3 x} - 3x + 2 = 0$
4. $3\log_2 x - 1 = \log_2 5$
5. $2 + \log_3(x-1) = \log_2 3$

Solution. Solve the following logarithmic equations:

1. $\log_2(3x) = 1$
   The equation is well-defined if and only if $3x > 0$. Then $x > 0$.
   We have
   $$\log_2(3x) = 1$$
   $$\log_2(3x) = \log_2 2$$
   $$3x = 2$$
   $$x = \frac{2}{3}.$$
   $\boxed{\text{Therefore, } x = \dfrac{2}{3}.}$

2. $\log_2 x + \log_2 4 = \log_2 5$
   The equation is well-defined if and only if $x > 0$.
   We have
   $$\log_2 x + \log_2 4 = \log_2 5$$
   $$\log_2 4x = \log_2 5$$
   $$4x = 5$$
   $$x = \frac{5}{4}.$$
   $\boxed{\text{Therefore, } x = \dfrac{5}{4}.}$

3. $3^{2\log_3 x} - 3x + 2 = 0$
   The equation is well-defined if and only if $x > 0$.
   We have
   $$3^{2\log_3 x} - 3x + 2 = 0$$
   $$3^{\log_3 x^2} - 3x + 2 = 0$$
   $$x^2 - 3x + 2 = 0$$
   $$(x-1)(x-2) = 0.$$

   It follows that $\left[\begin{array}{l} x - 1 = 0 \\ x - 2 = 0 \end{array}\right.$. Then $\left[\begin{array}{l} x = 1 \\ x = 2 \end{array}\right.$.

   Therefore, $x \in \{1, 2\}$.

4. $3\log_2 x - 1 = \log_2 5$
   The equation is well-defined if and only if $x > 0$.
   We have
   $$3\log_2 x - 1 = \log_2 5$$
   $$3\log_2 x = 1 + \log_2 5$$
   $$3\log_2 x = \log_2 2 + \log_2 5$$
   $$3\log_2 x = \log_2 10$$
   $$\log_2 x = \frac{1}{3}\log_2 10$$
   $$x = 10^{\frac{1}{3}}.$$

   Therefore, $x = \sqrt[3]{10}$.

5. $2 + \log_3(x - 1) = \log_2 3$
   The equation is well-defined if and only if $x - 1 > 0$. Then $x > 1$.
   We have
   $$2 + \log_3(x - 1) = \log_2 3$$
   $$\log_3 3^2 + \log_3(x - 1) = \log_2 3$$
   $$\log_3 9(x - 1) = \log_2 3$$
   $$3^{\log_2 3} = 9(x - 1)$$
   $$x - 1 = \frac{3^{\log_2 3}}{3^2}$$

$$x - 1 = 3^{\log_2 3 - 2}$$
$$x = 3^{\log_2 3 - 2} + 1.$$

Therefore, $x = 3^{\log_2 3 - 2} + 1.$

Problem 41. Solve the following logarithmic equations:

1. $\log_x 8 + \log_8 x = 2$
2. $2\log_x 2 + \log_2 x = 3$
3. $2\log_x 3 + \log_9 x = 2$
4. $\log_2 \left(x^2 + 4x - 3\right) = 1$
5. $\log_3 \left(x^2 + 4x + 1\right) = \log_3 2 + 1$

Solution. Solve the following logarithmic equations:

1. $\log_x 8 + \log_8 x = 2$
   The equation is well-defined if and only if $x > 0$ and $x \neq 1$. Then $x \in (0, 1) \cup (1, +\infty)$.
   We have
   $$\log_x 8 + \log_8 x = 2$$
   $$\frac{1}{\log_8 x} + \log_8 x - 2 = 0$$
   $$1 + \log_8^2 x - 2\log_8 x = 0$$
   $$\log_8^2 x - 2\log_8 x + 1 = 0$$
   $$(\log_8 x - 1)^2 = 0$$
   $$\log_8 x - 1 = 0$$
   $$\log_8 x = 1$$
   $$\log_8 x = \log_8 8$$
   $$x = 8.$$

   Therefore, $x = 8.$

2. $2\log_x 2 + \log_2 x = 3$
   The equation is well-defined if and only if $x > 0$ and $x \neq 1$.

Then $x \in (0,1) \cup (1, +\infty)$.
We have
$$2\log_x 2 + \log_2 x = 3$$
$$\frac{2}{\log_2 x} + \log_2 x = 3$$
$$2 + \log_2^2 x = 3\log_2 x$$
$$\log_2^2 x - 3\log_2 x + 2 = 0$$
$$(\log_2 x - 1)(\log_2 x - 2) = 0.$$

Then $\begin{bmatrix} \log_2 x - 1 = 0 \\ \log_2 x - 2 = 0 \end{bmatrix}$. It follows that $\begin{bmatrix} \log_2 x = 1 \\ \log_2 x = 2 \end{bmatrix}$.

- For $\log_2 x = 1$, we obtain $\log_2 x = \log_2 2$. Then $x = 2$.
- For $\log_2 x = 2$, we obtain $\log_2 x = \log_2 2^2$. Then $x = 4$.

Therefore, $x \in \{2, 4\}$.

3. $2\log_x 3 + \log_9 x = 2$
The equation is well-defined if and only if $x > 0$ and $x \neq 1$.
Then $x \in (0,1) \cup (1, +\infty)$.
We have
$$2\log_x 3 + \log_9 x = 2$$
$$\frac{2}{\log_3 x} + \log_{3^2} x = 2$$
$$\frac{2}{\log_3 x} + \frac{1}{2}\log_3 x = 2$$
$$4 + \log_3^2 x = 4\log_3 x$$
$$\log_3^2 x - 4\log_3 x + 4 = 0$$
$$(\log_3 x - 2)^2 = 0$$
$$\log_3 x - 2 = 0$$
$$\log_3 x = 2$$
$$\log_3 x = \log_3 3^2$$
$$x = 3^2$$
$$x = 9.$$

Therefore, $x = 9$.

Chapter 5. Solutions

4. $\log_2 (x^2 + 4x - 3) = 1$
The equation is well-defined if and only if $x^2 + 4x - 3 > 0$.
If $x^2 + 4x - 3 = 0$, we obtain the discriminant
$$\begin{aligned}\Delta' &= (b')^2 - ac \\ &= 2^2 - (1)(-3) \\ &= 4 + 3 \\ &= 7.\end{aligned}$$
Then
$$x_1 = \frac{-b' + \sqrt{\Delta'}}{a} = \frac{-2 + \sqrt{7}}{1} = -2 + \sqrt{7}$$
and
$$x_2 = \frac{-b' - \sqrt{\Delta'}}{a} = \frac{-2 - \sqrt{7}}{1} = -2 - \sqrt{7}.$$
It follows that $x < -2 - \sqrt{7}$ or $x > -2 + \sqrt{7}$.
We have
$$\begin{aligned}\log_2 (x^2 + 4x - 3) &= 1 \\ \log_2 (x^2 + 4x - 3) &= \log_2 2 \\ x^2 + 4x - 3 &= 2 \\ x^2 + 4x - 5 &= 0 \\ (x - 1)(x + 5) &= 0.\end{aligned}$$
Then $\begin{bmatrix} x - 1 = 0 \\ x + 5 = 0 \end{bmatrix}$. It implies that $\begin{bmatrix} x = 1 \\ x = -5 \end{bmatrix}$.

Therefore, $x \in \{-5, 1\}$.

5. $\log_3 (x^2 + 4x + 1) = \log_3 2 + 1$
The equation is well-defined if and only if $x^2 + 4x + 1 > 0$.
The discriminant of $x^2 + 4x + 1$ is equal to
$$\begin{aligned}\Delta' &= (b')^2 - ac \\ &= 2^2 - (1)(1) \\ &= 4 - 1 \\ &= 3.\end{aligned}$$
Then
$$x_1 = \frac{-b' + \sqrt{\Delta'}}{a} = \frac{-2 + \sqrt{3}}{1} = -2 + \sqrt{3}$$

and
$$x_2 = \frac{-b' - \sqrt{\Delta'}}{a} = \frac{-2 - \sqrt{3}}{1} = -2 - \sqrt{3}.$$
It follows that $x < -2 - \sqrt{3}$ or $x > -2 + \sqrt{3}$.
We have
$$\log_3\left(x^2 + 4x + 1\right) = \log_3 2 + 1$$
$$\log_3\left(x^2 + 4x + 1\right) = \log_3 2 + \log_3 3$$
$$\log_3\left(x^2 + 4x + 1\right) = \log_3 6$$
$$x^2 + 4x + 1 = 6$$
$$x^2 + 4x - 5 = 0$$
$$(x-1)(x+5) = 0.$$

Then $\left[\begin{array}{l} x - 1 = 0 \\ x + 5 = 0 \end{array}\right.$. We obtain $\left[\begin{array}{l} x = 1 \\ x = -5 \end{array}\right.$.

Therefore, $x \in \{-5, 1\}$.

Problem 42. Solve the following logarithmic inequalities:

1. $\log_2(2x) > 0$
2. $\log_3(x-1) > 1$
3. $\log_5(2x-3) < 1$
4. $\log_2(2x-1) \geq 1$
5. $\log_3(-x+1) \leq 2$

Solution. Solve the following logarithmic inequalities:

1. $\log_2(2x) > 0$
   The inequality is well-defined if and only if $2x > 0$.
   Then $x > 0$.
   We have
   $$\log_2(2x) > 0$$
   $$\log_2(2x) > \log_2 1$$
   $$2x > 1$$
   $$x > \frac{1}{2}.$$

   Therefore, $x \in \left(\frac{1}{2}, +\infty\right)$.

2. $\log_3(x-1) > 1$
   The inequality is well-defined if and only if $x - 1 > 0$.
   Then $x > 1$.
   We have
   $$\log_3(x-1) > 1$$
   $$\log_3(x-1) > \log_3 3$$
   $$x - 1 > 3$$
   $$x > 3 + 1 = 4.$$

   Therefore, $x \in (4, +\infty)$.

3. $\log_5(2x - 3) < 1$
   The inequality is well-defined if and only if $2x - 3 > 0$.
   Then $x > \dfrac{3}{2}$.
   We have
   $$\log_5(2x - 3) < 1$$
   $$\log_5(2x - 3) < \log_5 5$$
   $$2x - 3 < 5$$
   $$2x < 8$$
   $$x < \dfrac{8}{2}$$
   $$x < 4.$$

   Therefore, $x \in \left(\dfrac{3}{2}, 4\right)$.

4. $\log_2(2x - 1) \geq 1$
   The inequality is well-defined if and only if $2x - 1 > 0$.
   Then $x > \dfrac{1}{2}$.
   We have
   $$\log_2(2x - 1) \geq 1$$
   $$\log_2(2x - 1) \geq \log_2 2$$
   $$2x - 1 \geq 2$$
   $$2x \geq 3$$

$$x \geq \frac{3}{2}.$$

Therefore, $\left[\frac{3}{2}, +\infty\right)$.

5. $\log_3(-x+1) \leq 2$

   The inequality is well-defined if and only if $-x+1 > 0$.
   Then $x < 1$.
   We have
   $$\log_3(-x+1) \leq 2$$
   $$\log_3(-x+1) \leq \log_3 3^2$$
   $$-x+1 \leq 3^2$$
   $$x \geq 1-9 = -8.$$

   Therefore, $x \in [-8, 1)$.

www.ingramcontent.com/pod-product-compliance
Lightning Source LLC
Chambersburg PA
CBHW031631210526
45464CB00004B/1841